THE PACKAGE AS A MARKETING TOOL

THE PACKAGE AS A

MARKETING TOOL

STANLEY SACHAROW

CHILTON BOOK COMPANY
Radnor, Pennsylvania

For Beverly Lynn, Scott Hunter, and Brian Evan

PREFACE

Many people working in the packaging industry believe that the approach of "marketing the package" is unique to the last thirty years, an era of supermarkets and shopping centers. But enterprising companies have practiced this concept for more than a century.

Treating the package as the product probably reached its zenith during the heady, artistic days of the Art Nouveau period. Around the turn of the century, it was the package that raised a humble, ordinary food—the biscuit—into a position of importance on the tables of the Victorian diner. In England, the Quaker firm of Huntley & Palmers, and in France, the Lefevre-Utile bakery marketed scores of different tin biscuit packages in shapes ranging from train cars to biscuit barrels. Taking advantage of the artistic trends of the day, these and other pioneers were significantly ahead of their time and predated the birth of package design as a discrete profession. They developed their packages in a time before the term *demographics* was coined or the concept of a marketing mix created.

The advertising and marketing brilliance of these dynamic Victorian firms rapidly became known, and when the profession of package design finally emerged as a discipline in the late 1920s, the face of packaging slowly began to change. Catalyzed by the development of new materials, packaging took on the dual responsibilities of sales stimulation and consumer appeal.

With less Victorian "art" and more modern marketing "science," the package has proven to be one of the most important and viable advertising mediums available. This book treats the package itself as a product and shows the reader how to use it to increase sales. It covers all the elements of planning and executing a successful package, and explains how to make it an integral part of sound advertising, marketing and sales strategy.

CONTENTS

INTRODUCTION

The National Biscuit Company "did almost as much as the introduction of canned foods before it, and the invention of the electric refrigerator after it, to change the techniques of modern merchandising."

Fortune (April, 1935)

Until about 1900, a package had to perform only two functions: (1) to ensure safe delivery of the product to the consumer, and (2) to protect the product and ensure its shelf life. But at the turn of the century an entirely new dimension was added to the emerging field of packaging technology. The package itself became a sales agent: It had to be able to "protect what it sells and to sell what it protects."

As merchandising became more impersonal, manufacturers had to devise ways of distinguishing their product from those of their competitors. Although trademarks were used occasionally as early as the 1780s, the Smith Brothers actually pioneered the concept in 1866. The Smith Brothers put their cough drops into large glass jars characterized by the two brothers, William and Andrew, with the words "Trade" appearing above William and "Mark" above Andrew. This simple design prevented the competition from using similar-sounding names to confuse the customer.

Fig. 1 *Huntley & Palmers "Cavalry" tin, with hinged lid, c. 1894, 14.5 cm. by 15.3 cm. Lid: Royal Horse Artillery. Front: Royal Irish Lancers, Horse Guards, and 15th Kings Hussars. Sides: 3rd Prince of Wales Dragoon Guards and 6th Dragoons Carabiniere. Back: Scots Grey. Courtesy, Associated Biscuits Ltd., Kings Road, Reading, England.*

Although the idea of branding a product was not new, using the package to convey the brand name to the consumer was. The bulk goods that were transported to country stores from the major wholesale centers, such as New York and Chicago, had long been branded. Originally, a brand was a mark applied with a blacking brush or a hot iron to a bale or cask, showing where the contents came from or who shipped them. A brand on a container also meant that the contents had been examined and passed by a public inspector, so it also came to mean a grade or a certain quality. Buyers recognized the brand marks and began to rely on them.

In 1870, the first trademark was registered in the United States. It was the eagle of the Averill Chemical Paint Company. In England, the first registered trademark was the red triangle on Bass and Co's "Pale Ale," in 1876. Trademarks quickly began to reflect the times (see Table 1). By 1871, 121 trademarks were registered in the United States; in 1906 there were more than 10,000, and in 1980 more than 650,000.

TABLE 1   ANIMALS AS TRADEMARKS

Trademark	Company and Product
Cow	The Borden Company (dairy products)
Horse	White Horse Scotch
	Mobil Oil Co. (flying red horse)
Mule	20 Mule Team Borax
Camel	Camel Cigarettes
Lion	MGM and Leo the Lion (motion pictures)
Bull	Genuine Durham Smoking Tobacco (Bull Durham)
	Merrill Lynch, Pierce, Fenner and Smith
Cat (Chessie)	Chesapeake and Ohio Railroad
Dog	
(Nipper)	RCA
(Great Dane)	Great Dane Trailers, Inc.
(Greyhound)	Greyhound Bus Lines
Falcon	Eastern Airlines
	Ford Motor Co.
Peacock	NBC
Elk	Hartford Insurance
Kangaroo	D. Bumsted & Co. (salt sold in Victorian England)
Monkey	Toothpowder (India)
Turkey	Bell's Seasoning
Cougar	Mercury Automobiles (Ford Motor Co.)
Bull's Head	J&J Colman, Ltd.
Goose	Canadian Pacific Airlines
Goat	Great Northern Railway
Texas steer	
(Red River)	Symbol for Neiman-Marcus's private label line of clothing and food products
Alligator	Used as symbol for LaCoste sportswear
Moose	Moosehead Beer
	Moosabec sardines
Boar	Boar's Head provisions
Panther	General Motors
Mustang	Ford Motor Co.
Dog	Barking Dog pipe tobacco
Bear	Hamm's Beer
Lynx	Ford Motor Co.
Birds	Birds' Custard

Between the late 1860s and the 1880s, many factories started to pack their goods in small quantities wrapped in paper, and the consumer could ask for "a paper" of coffee or "a paper" of dried yeast. The papers were for convenience, and soon the wrappers began to advertise the contents. But perhaps it was the "Uneeda" package that really pioneered the idea that the package is a valuable marketing tool. Developed about 1895 by the National Biscuit Company (Nabisco), this novel paperboard carton not only kept the crackers fresher than the old cracker barrel, but also advertised the brand name.

Nabisco's "Uneeda" biscuit package marked the beginning of the colorful rows of products that line supermarket shelves today. It also signaled a new wave of marketing to capture brand loyalty.

In 1895 the C.W. Post Company of Battle Creek, Michigan, offered one-cent certificates on its new health food, "Grape Nuts." Thus the age of couponing had begun. And as early as 1907, the

Fig. 2 *The round paperboard carton of Quaker Oats, c. 1925. The ruddy-cheeked, pleasantly smiling Quaker man has long since become synonymous with the product that changed the nation's breakfast habits more than 100 years ago. Courtesy, Landor Associates, San Francisco: Museum of Packaging Antiquities.*

Fig. 3   *Tiger Tobacco tin of the P. Lorillard Co., 1895.* Courtesy, U.S. Tobacco Museum, Greenwich, Connecticut.

Franklin Sugar Company, pushing its boxed sugar, was able to point to a grocery store in New York that sold everything in packages, cartons, or cans, eliminating the time and effort devoted by traditional grocers to measuring and weighing whatever quantity the customer might ask for. "Join the Gathering Army," exhorted Franklin Sugar.

The dream of eternal prosperity that emerged in the 1920s disappeared with the stock market crash of 1929. Advertising budgets were drastically cut, and new, cheaper methods were needed to convey the sales message to the consumer.

The supermarket was born in mid-summer 1930, when Michael Cullen ran his first ad in Jamaica, Queens, New York: "Jell-O, 7 cents; Kellogg's Corn Flakes, 10 cents a box; Presto flour, 25 cents." Thus was born the King Kullen supermarket. "Pile it high, sell it low" was the motto. When customers entered, they were met by mountains of canned goods, stacks of packaged goods, pyramids of produce. Signs beckoned on every side:

"Save," "Save," "Save," "Save." The supermarket became an overnight success. "We couldn't get the stuff on the shelves fast enough" noted John B. Cullen, son of the founder.

The mid-1930s also saw the beginnings of industrial design. Its purpose: to capture the consumer's imagination. The package now had to play a dual role: It had to sell *and* advertise its contents.

Another legacy of the Depression years was the one-use container for beer, soft drinks, and milk. The beer can was introduced in 1935 by the American Can Company through the Krueger Brewery in Newark, New Jersey. In a 1936 article entitled "Beer into Cans" in *Fortune* magazine, the writer stated: "Realistic canmakers do not claim that the can is or ever will be cheaper than the bottle. The true advantages lay in shipping, in eliminating the need for the supermarket to deal with empty bottles, and

Fig. 4 *The shopping cart was invented in 1937. With self-service and a high sales volume, the supermarket could cut prices by 25 percent and set a new competitive standard for the grocery industry. Courtesy, Food Marketing Institute, Washington, D.C.*

in the large market for cans rather than bottles." Krueger offered two brands of canned beer, and within five months, the company was running 550 percent of its precan production.

During the Depression, the phrase "a package should entice its purchaser" often meant odd sizes, oversize packages, and increased attempts at customer deception. In the generation between the world wars packaging deception was adopted and institutionalized by the advertising agency.

World War II brought the need for more sophisticated production methods and decline of manual labor. But it was the postwar years that dramatically affected present-day packaging trends. The self-service concept spread, and the American supermarket became the worldwide symbol of plenty and variety. Shopping centers sprang up (probably the first was the Roland Park Shop Center in 1907 in Baltimore), and they became bigger and bigger. Shoppers' World in Boston in 1951 had more than 500,000 square feet of retailing space, while the original Garden State Plaza in Paramus, New Jersey, had 1,480,000 square feet. Even the supermarket has undergone change—from King Kullen's pioneering size to the megamarkets of the 1980s.

Labor costs rapidly increased. New factors such as an increase in the number of working women, two-income households, and single households caused more changes in packaging, and marketing strategies were modified to reflect these changes. Soon business discovered that the package could serve the consumer as well as the producer in almost every phase of marketing.

Affluence was a characteristic of the 1960s. Extravagant packaging was created to appeal to a better educated consumer group with more discretionary income. Convenience became an increasingly important factor in design. And toward the end of the 1960s, *ecology* became an important consideration in marketing.

With the passage of the Fair Packaging and Labeling Act in 1966, the age of consumerism reached maturity. More and more people wanted to know more and more information about the product and how much it really weighed. Nutritional labeling, generic foods, "no-frills" labeling, the changeover to the metric system, and child-resistance have all become important packaging considerations. The prospects for a national beverage container law or even a tax on all packaging to be passed in the 1980s is strong. Package variations are being slowly reduced because of both cost and availability. Shortages in natural resources eventually become evident in higher costs and minimal availability of many plastics, papers, and metals.

The 1980s will also bring a shift in demographics as the "baby boom" children approach middle age. And the later decades of this century promise even greater packaging innovations, as we shall see. More women will be working, with less time available for preparing food. Convenience items promise to capture even greater market segments than they enjoy today. Together with the steadily increasing "mature market," the 1980s promise to be an exciting decade for the packaging industry.

1 DEMOGRAPHICS AND THE AMERICAN CONSUMER

Consumers, by definition, include us all. They are the largest economic group in the economy, affecting and affected by almost every public and private economic decision. Two-thirds of all spending in the economy is by consumers.

President John F. Kennedy

Different markets often require different packages. Just as the converter prices the same material according to the market, the resourceful product manufacturer knows that the package that sells for one group might fail for another. For example, because of consumer preference, potato chips are usually sold in transparent packages on the East Coast and in opaque bags on the West Coast. This is because opaque bags were first introduced on the West Coast and consumers never changed their initial preference. The failure of Mattel to capture the European market with its successful "Barbie" doll was partially because the company did not consider the difference between European and American tastes in clothes. And while food in tubes sells briskly in European supermarkets, the average American consumer associates the tubes with toothpaste.

A sound marketing program begins with a careful analysis of consumer demand for the product or service. A company that is production-oriented may treat its entire market as a single undifferentiated unit. Thus a product intended for one market will be aimed at reaching as many consumers as possible. A small converter of packaging materials may produce a single laminate and attempt to sell it to whomever possible. No specific market is intended and the material can be used to package a wide variety of products. The basic aim of the firm is to "push the product" out as rapidly as possible. When competition becomes severe, a firm may try to differentiate its products from those of its competitors, perhaps by changing the package or introducing a new size, flavor or material. But the total consumer market is too diverse to consider as a whole: There are simply too many consumers with too many differences. Thus package design must be considered in terms of demographics and trends in order to be successful. One reason for the projected success of the retort pouch (a shelf-stable, flexible pouch for food) is that it offers superior quality and convenience for the person eating alone. This is important because single-person households are growing even faster than two-person households; census figures show that households of people living alone are increasing three times as fast as family units. Further, fewer meals are being eaten at home, mainly because of the convenience. Just under 40 percent of each dollar spent on food goes for meals eaten outside the home.

Statistics recently compiled by the United Nations Fund for Population Activities note that the world population is expected to increase from 4.3 billion in 1980 to 6.2 billion by the year 2000. In 1960, 115 cities in the world had populations of more than one million, or 10 percent of the population. By the turn of the century, there will be 440 such cities, accounting for 22 percent of the world's population. The five largest cities in the year 2000 are expected to be Mexico City, 32 million; Tokyo-Yokohama, 26 million; São Paulo, Brazil, 26 million; New York-northeast New Jersey, 22 million; Calcutta, 20 million.

The population of the United States is also increasing rapidly. By the year 2000, the population will exceed 300 million, up from 100 million in 1915. Coupled with a declining birthrate, the trend toward a higher standard of living and greater labor productivity becomes pronounced. The average household size is also projected to decrease by 1990, and this will mean an increased demand for smaller packages and more variety in product type. This vast consumer purchasing power can be reached only by segmenting the population into discrete markets. The power of the package can often be used to its fullest advantage only when integrated with a scheme of market segmentation.

REGIONAL DISTRIBUTION

Although the biggest markets now are in the East and Midwest, the greatest percentage increases in the 1980s are expected to be in the Sun Belt—the southern and western regions. This means more bright, warm colors in package design and an increased demand for outdoor foods, such as barbeque sauce and frankfurters. Demand will increase for suntan lotion, swimsuits, backyard swimming pools and other warm-weather products. People in the West are less formal than Easterners and they spend more time outdoors. As the Western population increases, there will be a larger market for items such as patio furniture, sports clothes, horse riding equipment and outdoor grills.

While bright, warm colors are preferred by people in Florida and the Southwest, grays and cooler colors predominate in New England and the Midwest. Products intended for regional distribution in these areas should take these color preferences into account.

Many people in the South prefer to take "headache powders" (aspirin) in the form of a powder neatly folded in a glassine wrapper. Drug manufacturers with southern plants cater to this regional peculiarity by producing the package on modern high speed machinery. In other parts of the nation, this type of product-package is almost unknown. Aspirins are invariably taken as tablets and not in the form of a fine powder.

Regional differences are broken down by cities, suburbs, and rural communities. For example, suburbia has long been the traditional area for fast-food marketing. But with changing demographics, fast-food outlets in metropolitan areas are expected to increase. Workers in Manhattan usually go out for lunch and thus are potential fast-food customers. People can walk to a fast-food outlet and not be concerned about a gasoline shortage. People in the suburbs also tend to eat at home more often than city dwellers. More than 85 percent of all fast-food customers live or work within 2½ miles of the outlets they patronize. This means that whole new markets for fast-food are about to open in metropolitan areas.

TABLE 1.1   TOP TEN FAST-FOOD COMPANIES
ANNUAL SALES (MILLIONS), 1978

	Domestic Sales	Percentage of U.S. Market
McDonald's	$3,848.0	18.2
Burger King	1,130.7	5.4
Kentucky Fried Chicken	1,200.0	5.7
Wendy's International	783.4	3.7
International Dairy Queen	823.0	3.9
Pizza Hut	724.0	3.4
Hardee's	460.8	2.2
Sambo's	538.6	2.6
Denny's	414.0	2.0
Jack in the Box	430.0	2.0
Total Top 10	$10,352.5	49.1
Total Industry	$21,100.8	100.0

Source: *Advertising Age* (September 22, 1980).

DISTRIBUTION BY AGE

In the 1980s, the age mix will reflect not only the low birthrate of the 1930s but also the baby boom of the post–World War II years. (*U.S. News & World Report* estimates that the age 35–49 population will grow by 76 percent by the year 2000, with an increase of 53 percent in population over age 50.) Complicating the market even further is the declining birthrate of the 1970s. This means that in the 1980s and well into the year 2000, the population will grow at a slower rate, whereas the aging population will increase. Trade sources project a 42-percent increase during the next decade in the number of Americans aged 35 to 44. As the baby-boom generation approaches midlife, the emphasis will be less on youth. The huge youth market will become the booming young-adult market, and typically it will have a far different set of values and lifestyles than the previous generation.

The youth market (ages 5 to 13) often influences parental purchases; indeed, billions of dollars are spent on this group by their parents. In addition, this group often makes its own purchases of goods and services. Children's television shows are often sponsored by cereal manufacturers and other advertisers in an effort to develop brand preferences at an early age.

The teenage market, perhaps the one that exerts the most significant influence on the methods used to market products in this country, is the most difficult to reach, even though it is a proven market for snacks, convenience foods, and soft drinks. A study conducted by the Coca-Cola Company in 1977 found that people in the 16-to-19-year age group tend to have the highest per capita consumption of soft drinks. Soft-drink packaging *must* be aimed at this group. With an increasing amount of money to spend, teenagers are good customers for records, automobiles, cosmetics, clothes, jewelry and other products. Many manufacturers are adopting new product and distribution policies. Some clothing manufacturers are now designing junior ready-to-wear dresses which reflect the age and not merely the size of the teenager.

There are instances where the teenage market has proven to be difficult to reach. While some companies have had success, others have failed in their attempts to reach the older end of the youth market. Many firms tend to lump all teenagers together in one group, instead of considering the many subgroups based on income, race and geographic location. Certainly the 14-to-16 age group is different from the 17-to-20 group.

Recently, more firms have been taking a closer look at the large and growing, financially secure maturity (45 to 64) market. There are now about 56 million Americans age 50 or older, about 26 percent of the total population; more than half the total households in the country, about 39 million, are headed by someone 45 or older. This group usually eats at home, rarely skips meals, and controls 60 percent of all discretionary income. Older Americans spend a greater percentage of their income (22 percent) on food than those under 65 (17 percent) and more per person on food at home than any other age group.

With improved health care, a dwindling birthrate, and the aging of the post–World War II baby boom, the maturity market will continue to increase. In fifty years, an Ogilvy and Mather forecast projects, there will be just as many people age 45 to 64 (78 million) as there are people 18 to 34.

Dietary supplements, special foods, and health care items are special products for this market. Products designed for and aimed at younger markets will become saleable to the more mature (see Table 1.2). The packages also will have to appeal to those over 45. Johnson and Johnson is now persuading adults to use its baby oil and baby shampoo. Levi Strauss markets a three-piece suit and "Levis for Men" that are more fully cut to accommodate the man who has stopped playing football and is now watching it. And Del Monte expects that sales of prune juice will go up as people grow older.

Of course, no product or package should be advertised by age. The failure of Heinz's "Senior Foods" is a notable case of advertiser's error. Seniors failed to purchase the product and the entire line failed in the marketplace. Even when the product is likely to

Fig. 1.1   *This paperboard carton, c. 1925, is an early attempt at package design specifically aimed at the 18-to-22-year-old college market. It held various cosmetic items.* Courtesy, collection of Edward Morrill, Werbin & Morrill, Inc., New York City.

be of particular interest to seniors, experts advise that the stress should be on the product, not on the user. The percentage of people over 55 who have hypertension and need to restrict the salt in their diet is tremendous. This means that food manufacturers should start introducing a wide range of salt-free products intended for both younger and older consumers.

TABLE 1.2   NEW FOOD PRODUCTS AIMED AT MATURE MARKETS, 1980

Manufacturer	Product
Kellogg	"Smart Start" bran cereal
Proctor & Gamble	"High Point" instant decaffeinated coffee
Anderson-Clayton Foods	"New Age" low-cholesterol cheese
Wm. Wrigley Jr. Co.	"Freedent" gum (does not stick to dentures)
Kraft Foods	"A la Carte" individual-serving entrées

DISTRIBUTION BY SEX

Market segmentation by sex is important because certain products are purchased more by one sex than another. Not too many years ago, the wife did practically all the grocery shopping for her family, whereas the husband bought the products and services needed for the automobile. Today, more men are food shoppers and more women buy the gas and service the car.

In addition, the changing role of women can be seen in the increased number of working women, married or single. Women accounted for 32.3 percent of the labor force in 1960, 36.8 percent in 1970, and 38.5 percent in 1980. Working women have different shopping patterns than nonworking wives. Over the past several years, a working wife has contributed about 25 percent of the total family income. In those families where the wife works, median family income is substantially higher than in families where just one member works. This makes more discretionary money available to the family for purchasing products other than necessities or for purchasing items of a higher quality.

Better packaging always emphasizes higher quality products. Also, better packaging will attract attention to discretionary items. Since many discretionary items are bought on impulse, attractive and appealing packaging is of great value in stimulating sales. People pay more for products if the packages are psychologically and aesthetically appealing.

The important role of the working woman in today's society has also caused many advertising agencies to study its effect on spouses. A 1980 report, issued by Cunningham and Walsh and summarized in the *New York Times* (December 15, 1980), stated, "Three years ago everyone was interested in the working woman: she represented such a huge market. Now, agencies are realizing that the lives of husbands had to change, too. Ads for beer have always been telecast during ballgames; maybe we should be showing soap flake commercials during these times."

FAMILY LIFE CYCLE

Markets are also segmented by the various stages that a family undergoes in the life cycle. Single households tend to be more affluent, more mobile, more oriented to immediate pleasures, and more interested in leisure-time activities and fashions than family households.

There are some other interesting demographic trends in the life-cycle category. About 55 percent of all households have no more than two people, compared with 40 percent in 1950 and 46 percent in 1960. Divorces rose 55 percent between 1970 and 1977, and single people now comprise 30 percent of the population, compared to only 15 percent in 1960. Single-parent homes increased 35 percent between 1970 and 1977, and childless families increased by 17 percent. Between 1965 and 1974 the number of couples who live together has increased more than 800 percent.

ETHNIC GROUPS

For some products, it is often useful to analyze population on the basis of race, religion, or national origin. One ethnic group that has received a substantial amount of attention is the growing Black market, which contains an estimated 25 million consumers with a combined buying power of close to $100 billion. Health and

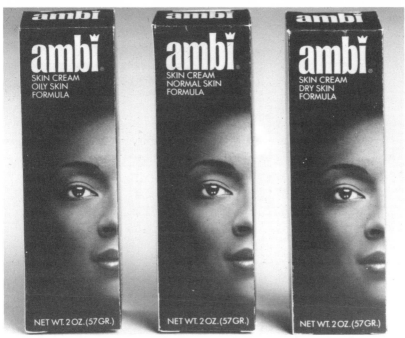

Fig. 1.2　*The Ambi cosmetic carton has been designed to appeal to the young Black woman. Courtesy, David Deutsch Advertising; designer, David Deutsch; photographer, Rosemary Howard.*

beauty aids alone account for about $1 billion. The annual growth rate of this market is $100 million. There are definite characteristics in the Black market:

1. Spending patterns are different than for whites, given the same income.
2. Black consumers tend to be loyal to a product line.
3. Blacks are considerably younger, with a median age of 21 against the white median age of 29.
4. Blacks will account for one-fifth of the net population gain between 1980 and 1990.
5. Black tastes reflect Southern origins. Black consumption of soft drinks in Chicago follows the same pattern as in the South. Colas, grape, and orange drinks are favored.
6. Two-thirds of Blacks live within the inner city. Smaller packages appeal to them since they cannot afford large size purchases.

7. Blacks tend to spend more of their disposable income on tangible items, rather than on intangibles.
8. Blacks spend 23 percent more per capita for shoes than whites.
9. Blacks spend up to 12 percent more per capita on food than whites.

The amount of space devoted to Black health and beauty aids depends on the type of store and its volume. However, in general, if 10 to 25 percent of the customers are Black, four to eight feet of space should be alloted to specialty products. If 25 to 50 percent are Black, 8 to 12 feet is appropriate, while 50 to 75 percent warrants 12 to 16 feet. If more than 75 percent of store traffic is Black, then more than 20 feet of space should be devoted to items such as Black health and beauty aid products.

Many retailers find ethnic products difficult to evaluate— several new products may be introduced in any given month— and they should rely on specialty jobbers who keep up with the latest trends.

Another ethnic group of increasing importance is the 20 million Hispanics (41 million by the year 2000), of which 7 million are illegal aliens. Spending more than $30 billion with a gross national income of $51.8 billion, the Hispanic population in America

Fig. 1.3 *.Toiletries for the Black market.* Courtesy, *Day Laboratories, Inc., Milwaukee, Wisconsin.*

by 1985, according to government estimates, will exceed that of any other minority. Just as with the Black market, there are certain characteristics of the Hispanic consumer (a recent study, *U.S. Hispanics —A Market Profile* is available from its publisher, Strategy Research Corporation, Miami, for $50):

1. The Hispanic market has experienced a growth rate of 124 percent in 15 years.
2. The Hispanic population is concentrated in California, Texas, New York, New Mexico, Arizona, Florida, Colorado, and Illinois. In the Greater New York Area, there are 2.3 million (up from 991,900 in 1970).
3. The median age of the Hispanic consumer is young—21 to 22 years old.
4. The median annual income for Hispanic households in 1979 was about $14,000.
5. Hispanic consumers tend to buy the best they can afford and are conservative and family oriented.

ECONOMIC CLASS

The grocery costs in stores located in middle-, and upper-income neighborhoods are lower than in ghetto areas. This cost differential often ranges from 18 percent to a whopping 60 percent, and there are many reasons for this difference.

Small stores often stock only the smallest or largest sizes of common household items, such as bleach or detergent. And they must charge more for merchandise, since they cannot buy in bulk or count on fast turnover. Markets in low-income areas typically sell more sugar in two-pound bags than in five-pound bags and more milk in quart containers than in the relatively less expensive half-gallon containers. Often spaghetti, peanut butter, and pork and beans are priced higher in those neighborhoods where they might be expected to be a steady part of the diet. This often works against the poor consumer.

In addition to price, many low-income area stores are sloppy and unkempt. Packaged goods are often mismarked, frozen foods half-thawed, and meat that has been on display for several days often repackaged, relabeled, and redated.

Many other factors should be considered in a complete demographic study of consumer habits. People alone do not make a market; they must have money to spend. A detailed study of income, its distribution, and how it is spent is also essential in any quantitative market analysis. The distribution of income affects the market for many products. The income pyramid has been turned upside down over a period of twenty to thirty years. Family expenditure patterns vary depending upon family income and the stage of the family life cycle. Families with older children spend relatively large amounts on foods. Younger families must devote large sums to buying and furnishing a house. Other demographic shifts must be taken into account in the 1980s:

1. More working women (over 50 percent of all families)
2. Six million fewer children under 13 than in 1970
3. Five million more Americans over 65
4. Rapidly growing 25-to-34-year-old population
5. Rapidly growing Black and Hispanic markets

People are living in an era of rapid and far-reaching change. Never in the history of the United States has change been so extensive in all walks of life. Not only has the overall outlook of the population changed, but beliefs in technology and values have also changed. At the present time, there is every reason to believe that these changes will continue to take place in the future, at perhaps an even faster pace.

FOREIGN MARKETS

Although American consumer habits are probably the most complex in the world, the consumers of other nations often exhibit many similar demographic breakdowns. Denis Thomas in *The Visible Persuaders* (Hutchinson of London, 1967) reports that in Great Britain, households in London and the southeast spend proportionately less on personal goods such as alcohol, clothes, shoes, and cigarettes than those in the Midlands, Wales, northern England, or Scotland. They also spend more than any other region on services (telephones, hairdressing, laundry, education, vacations). Families with less money to spend seem to economize, not by buying cheaper varieties of products, but by going without more often. In the Midlands and east, more is spent on car travel than in other areas. In the south and southeast, less money is spent on food, clothes, and tobacco.

In Wales, the largest savings are on services and household durables. In the north they save not only on services but also on motor travel. In Scotland, they spend proportionately more than anybody on shoes and tobacco. In London and the southeast, families spend 25 percent more than the national average on poultry, mutton and lamb, fresh green vegetables, and fresh fruit. In the north, they spend 25 percent less than the national average on fresh green vegetables, but 25 percent more on margarine. In Scotland, they spend less on bacon and ham, flour, mutton and lamb, poultry, fresh green vegetables, and pork, but more than any other region on cakes and biscuits. The Midlands spends more on tea, baked beans, breakfast cereals, and sugar than other regions, and Scotland spends 25 percent less than the national average on tea, coffee, flour, breakfast cereals, frozen foods, and marmalade.

The British tend to eat breakfast foods at other meals. There is still no mass market in England for wine. The English do not eat as much soup as the Scots (who consume four times as much) or the French (ten times as much). Only one in twenty Englishmen will switch from their favorite brand of cigarettes.

Different nations often have varying consumer habits. Germans prefer peppermint toothpaste, whereas the British prefer spearmint. One of the most successful toothpaste packages in Italy is Durban's, which features old-fashioned pictures of dentists or toothaches. This design would probably fail miserably in the United States because Americans prefer clean, scientific and modern-looking package designs.

About 40 percent of West German women use lipstick, while 75 percent of English women do. Only 6 percent of Englishmen use any kind of toilet water, while 12 percent of the Dutchmen do and 69 percent of Frenchmen do.

2   MANAGING THE PACKAGING FUNCTION

With $25 billion annually being spent on packaging (about 8 percent of the total value of goods produced), there is good cause for a company's secretary-treasurer, comptroller, or vice-president of finance to take a hard look at the packaging function.

Walter P. Margulies, *Packaging Power* (1970)

Only since the mid-1950s has American business recognized packaging as an essential part of its commerce. The distribution of consumer wealth created a new era in which all consumers had the means to choose their purchases freely. It was in these years that rapidly increasing government legislation of food and drug additives led many in American industry to realize that by properly managing packaging activities, company profits could actually increase.

Suddenly firms began to organize packaging departments, hire coordinators, and scout outside consultants in an effort to quantify their packaging decisions. Although many were unaware of the new packaging technology, they were willing to adapt and learn on the job. In 1956 the first class of packaging engineers was graduated, from Michigan State. Soon, a new, bright, energetic cadre of packaging professionals appeared on the scene, but often they were forced to report to superiors who had only a passing knowledge of the field. While this problem still exists, the last decade has seen a more organized packaging function arise, one that was not subordinate to another department but capable of standing on its own.

Because packaging is a service function, it is often difficult to measure its effectiveness. When it is effective, it benefits the business. When it is not, it can cause problems and sap profits. Management can profit by understanding the key areas of the packaging function and assuring that these are handled effectively.

The packaging function can be managed in many ways. A product manufacturer looks at his packages somewhat differently than the material supplier who supplies the material for the package. Even in the product manufacturer's organization, a certain degree of variation exists. Four basic techniques are used by product companies to manage the packaging function:

1. Full responsibility is assigned to one department (see Fig. 2-1)
2. A packaging committee of three to five people from various departments is organized
3. A small business might assign the function to one person (packaging coordinator and/or packaging consultant)
4. In special cases, a source outside the company, such as a contract packager, would be enlisted

The packaging function also can vary from company to company, since it covers such a diverse range of disciplines, including:

> Packaging development
> Package engineering
> Production and production support
> Quality control/quality assurance
> Marketing
> Package design (functional and aesthetic)
> Legal (must meet all legislative requirements)
> Medical (conforms to any applicable stipulations on content, warnings, etc.)

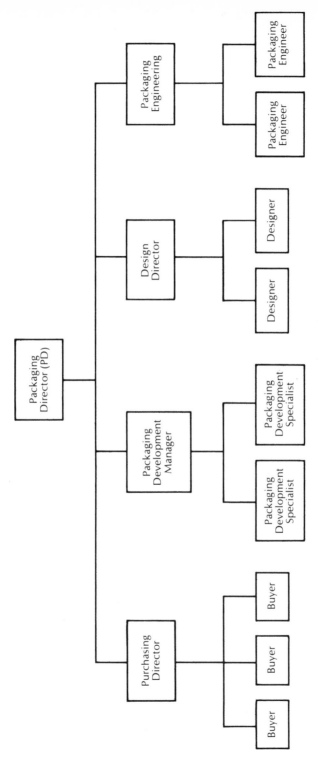

Fig. 2.1 *Representative packaging department.*

Purchasing (equipment and packaging materials)
Product/Pack stability
Warehousing and distribution

The power center for the packaging operation often depends on personalities (marketing personnel frequently consider their role to be "controlling," since within the company they are the first customers), or on a particular emphasis within the company. A company with a relatively few but major selling lines may be motivated by output and production efficiency; "control" may reside in the production area. However, all activities contribute to the whole; communication and coordination are essential to reduce any barriers that might be motivated by internal competition.

The total activity is probably best controlled under the umbrella of an experienced packaging coordinator, vice-president, or packaging director. Coordination always must involve:

People
Materials
Documentation (protocols, standard operation procedures, records and reports)
Facilities and equipment
Costs

THE PRODUCT MANUFACTURER

The manufacturer must check the mode of shipping to ensure safe delivery to the retail outlet. The caliber of paperboard used in a carton must be able to withstand shipment and seals on plastic bags should not break on the retail shelf. Retailers and manufacturers must work as a team. A product must be stacked and displayed properly on a store shelf. Items must not be subject to excessive damage through improper storage. Smudged printing, easily broken bags and crushable cartons will put the consumer off and lead him or her to a competitor's product. A consumer's initial reaction to a product is through its package and a poor package often spells disaster for its producer.

While the manufacturer's basic concern is his product, the package must be as carefully developed as the product itself. Items such as shelf space, lighting, typography, etc., must be taken into account. This calls for the integrated teamwork of people from different disciplines—all willing to contribute to the market success of a new product.

Most medium-to-large producer firms have formal packaging departments. These vary, depending on the size of the company, from two or three persons to multisectioned departments, such as Johnson and Johnson has to coordinate its many successful drug and cosmetic items.

The Packaging Director (PD)

The packaging director must have an intimate knowledge of the industry. He holds a top executive position and has full responsibility for all facets of the package's integration into the product manufacturing process. In a large organization, the PD must be aware of all facets of the operation, with an obvious slant toward packaging.

Fig. 2.2 *Marketing the packaged product.*

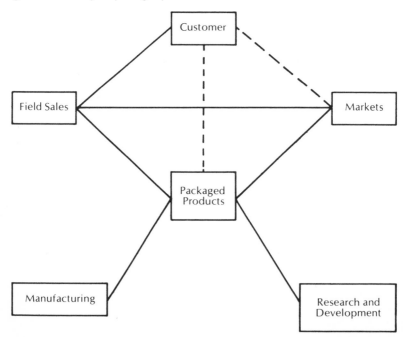

Perhaps the best training for the PD is to have worked either in packaging purchasing, research, or manufacturing. Having worked in all three is even better because the PD must be a generalist, not a specialist, and be aware of current FDA legislation and familiar with ways to meet escalating packaging material costs.

One of the most important departments that report to the PD is purchasing. The purchasing department's responsibilities are:

1. To assure suppliers that ideas will be evaluated by others as well as the buyer
2. To establish communications within the firm to make certain that packaging problems and complaints are referred to Purchasing and dealt with promptly, and that packaging criteria are furnished by marketing and sales
3. To recognize commitments and obligations to suppliers, place values on service, quality, and past relationships, and make suppliers aware of policies and procedures relating to the procurement of present supply needs and newly developed packages

From specifications derived by other departments under the PD, purchasing's job is to provide advice and service in the acquisition of specified packaging for all departments of the division and to maintain desired packaging inventory levels. It's not an easy job. Purchasing must locate, develop, and select sources of supply. All standards imposed on the material by the specifications are also generally included in the purchase contract. Purchasing is also responsible for all other contract elements, such as quantities, price, and logistics.

Most purchasing departments are headed by a purchasing manager and a staff of buyers or purchasing agents. Often the only person a material supplier meets in the user firm is the buyer of the material being sold. Large firms such as Eastman Kodak or Warner-Lambert have highly organized purchasing departments, each with a discrete function.

Professional purchasing personnel often aid the material supplier. They usually serve as the clearinghouse for salesmen and direct his inquiry to the right person in the user organization. Bypassing a user's purchasing agent or buyer is usually unwise because when the time comes for packaging material purchase, no relationship has been established between the supplier and the person assigned the job of purchase. Price haggling and possible business losses may result. A more personal relationship, established earlier, often paves the way for success later on.

There appears to be a trend toward placing full responsibility for package development on the purchasing department. In several large companies, packaging development is largely under the jurisdiction of the purchasing department, although a short time ago responsibility was shared by the research department. Organizationally, packaging presents difficult problems because it involves almost every department, including the credit department.

Another department reporting to the PD is the Package Development Group. Composed of trained packaging technologists, this department oversees technical problems relative to the packaging material. It also must formalize packaging specifications, standards, and test methods. Individuals working in this department are usually trained scientists (chemists), engineers, or better, *package* engineers. They deal with the material suppliers (through purchasing) as well as with their own internal departments such as marketing, purchasing, and sales. Therefore, an essential prerequisite for success in this job is the ability to communicate effectively.

At present, the packaging community is divided into those concerned with the technical aspects of the field and those dealing only with graphics and aesthetics. No package could be sold without the services of the package designer: Here the artist meets the consumer. The individuals in this department are often commercial artists who have limited knowledge of the packaging process. They report to the PD, who often has little or no knowledge of the designers' activities. Thus it is essential that the designer be integrated in the total system. He or she should be fully aware of the needs of the market and design around these parameters. Often, this is difficult since the designer lacks the technical and sales training needed by the other personnel.

The 1970s have seen the emergence of the package designer as market researcher, a catch-all designation that covers a variety of functions not undertaken by the designer. Package designers now arrange for or conduct consumer testing. In one case, a package design firm worked with the producer to make sure the color of the product was just the right shade to appeal to the customer.

There are many advantages for a product manufacturer in having its own staff designer. Although suppliers often provide free design services to a good customer, their basic objective is to design packages to sell their materials. This limits the materials and processes they are able to recommend. If the user understands this limitation, no problems occur.

Finally, but not least in importance, is the package engineer. Trained in both mechanics and materials, he must guide the mechanics on the line and meet a host of machine problems head on. His job includes the running of new materials, recommendation of new packaging equipment, cost justifications of new equipment purchases, and supervision of machine start-up and installation. It is an essential function for any firm. One problem area in this job is that the package engineer must keep up with new trends and travel to suppliers on a regular basis. If his knowledge is limited to his firm only, he will not know how to make future production improvements.

The Packaging Committee

In many small-to-medium user firms, packaging decisions are often made either by the boss's wife or by a packaging committee because they cannot justify the large expenditure necessary to organize and maintain a separate packaging department.

The committee method is widely used and operates with representatives of all the areas most concerned with packaging. Members of the committee might include production, engineering, research, and procurement. Meeting on a regular basis, the committee ensures an effective flow of communication among all areas and fosters good-will among the separate departments. The important thing is to have a system under which all departments are consulted on packaging problems. No individual is in full charge of packaging. This situation calls for a high degree of cooperation in the internal organization as well as between the product manufacturer and sources of supply.

The packaging committee serves a useful and important function. Even a capable packaging engineer might not consider that screw caps of the same size on different size glass jars of an expensive food product, the price of which is stamped on the cap, may invite unscrupulous customers to switch caps to get the larger size at the lower price. The sales department might be aware of this and warn the packaging engineer. Because of this preventive function, it is vital that each department be given the opportunity to forestall mistakes that may be of concern to that

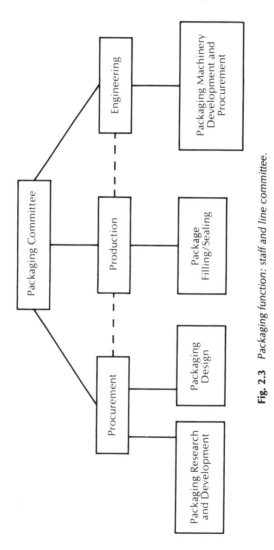

Fig. 2.3   *Packaging function: staff and line committee.*

department. A packaging committee does not create a good package, but it may prevent a deficient package from going into production (see Fig. 2-3).

There is still another approach often used by product manufacturers: A packaging coordinator becomes responsible for assimilating all the necessary information from brand supervision, engineering, and production. One problem with this approach however, is that a coordinator often lacks authority to make the job truly responsible. For example, a packaging consultant may provide a specialized knowledge of overall packaging considerations which few individuals within any company can either understand or afford to put into action.

The senior marketing executive must decide which approach to use. But no single person is capable of bearing the burden of making the final packaging decision. The best packaging management is not a simple caretaker operation. It must be geared to change, be creative and innovative, and oriented to facts and figures.

THE MATERIAL SUPPLIER

Along with the growth of market-oriented selling has come a realization by packaging-material suppliers that they must fully understand the industry they are attempting to sell. More sophistication by product manufacturers has also created the need for people in the supplier firms that can speak their language and relate to the suppliers' customers. This has led to the growth of technical organizations alongside the commercial, marketing, and sales departments within the suppliers firm.

The Sales Department

The sales department is usually headed by a national sales manager. On the sales map, the United States is divided into four or five geographical regions and each region is under the direction of a regional sales manager. Each region, in turn, is subdivided into as many as eight districts, each headed by a district sales manager. Sales representatives report to the district sales managers.

Fig. 2.4 *Garwood's Gloweave shirt box is a vacuum-formed blister following the contours of the shirt, thereby eliminating the need for pins and cardboard. The pack is vacuum-formed in two parts: Base material is white or clear .020" PVC with the lid in clear PVC; the lid is hot-foil stamped. The pack can be opened or closed quite easily; it remains rigid when closed and will not fall open. A novel and useful retail pack; winner of the PCA Development Award (Australia, 1980). Courtesy, Garwood Packaging Pty. Ltd., Bayswater, Victoria.*

Fig. 2.5 *This novel Dutch steam iron unit makes a good product presentation and is also economical to fabricate. Courtesy, Philips Persdienst, Eindhoven, The Netherlands; photograph, Fotostudio Peter Voorhuis.*

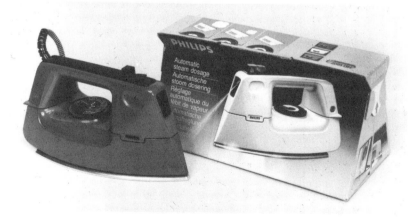

In addition to sales representatives and their managers, some of the larger regions might have one or more technical experts who also report to the regional general manager. These include package designers, artists, engineers, account executives, and regional product managers. Each function contributes services to the solution of the customer's problems and, in general, acts as a support force for the sales representatives.

Also in sales are the administrative or sales service personnel. This group may operate through a regional administration manager or a regional vice-president. This is logical, since their services cross all divisional lines, providing the vital inside sales functions necessary to support outside sales activities.

The Marketing Department

The marketing function is the responsibility of a national marketing manager. This is an all-important function, since marketing is the master discipline under which the four "Ps" are placed: *Product, Price, Promotion,* and *Place.* Reporting to the marketing manager are about six marketing directors and a market planning director. The marketing directors are responsible, through market managers, for the major market areas served by the user firm.

The key function of the individual market specialist is to develop programs aimed at a specific customer group, such as the beverage industry, the dry foods industry, or the meat industry. Program materials supplied by this division include all information necessary to introduce new packaging materials. The market managers also work closely with the advertising department to determine the type and scope of advertisements to use, such as newspapers, magazines, radio, TV, and trade shows. They also assist in the preparation of slide presentations and films and arrange traveling exhibits.

Other responsibilities of marketing people include the handling of various packaging exhibits and seminars, assistance in plant tours, and working directly with sales representatives in calling on product manufacturers.

Fig. 2.6   *The difficulty of wrapping rolls of steel strapping band has been solved by this Dutch development: a slit, continuous paper strip is used for circular wrapping. Courtesy, Meuwissen Industrie BV, Haarlem, The Netherlands; photograph, Fotostudio Peter Voorhuis.*

Administration and Control

The Administration and Control group, headed by the A&C manager, consists of three sections: Budget and Forecast, the Division Controller and his staff, and Pricing, Inventory, and Production Control.

In addition to preparing budgets and forecasts for the supplier, the manager also furnishes and analyzes the many statistics required for realistic short- and long-range planning. The controller, on the other hand, is responsible for the suppliers, adhering to approved budgets, and for all other financial transactions, including periodic reports to management and to the corporate controller. Another function of the controller is the establishment of and control over accounting systems in the supplier's manufacturing plants.

The most important function of the Pricing, Inventory, and Production Control director is to see that product pricing ensures an adequate financial return on investment. Product and raw-material inventories must be maintained to meet anticipated demands without being unreasonably large. The third function, production control, involves predicting and controlling the rate at which the many products are produced, to assure the most profitable product mix. Many products may be produced at more than one plant, so production control schedulers must decide where and when they can be produced most efficiently.

Product Management and Technical Services

Product Management and Technical Services, the fifth department reporting to the General Manager, is made up of product specialists and technical experts who support the efforts of sales representatives. Product Management, is divided into six or more categories, each under the direction of a product manager who reports to the Products Director. The major function of Product Management is to determine the proper packaging materials for specific applications. It works with both distributor and field sales people to help secure new customers and maintain existing accounts, in addition to serving as working links between sales and manufacturing and between sales and research.

There are two basic types of product or brand managers. The "General Foods" type is a generalist who knows a little about everything, similar to a general manager of his own business. The "Proctor & Gamble" type is a specialist in promotion, actively involved in all aspects of consumer relations; he also works closely with advertising agencies.

Technical services is responsible for new product planning and assistance to sales. Problems with field accounts are handled and new areas of possible technical interest to the company are explored.

Advertising

Although not an integral part of the suppliers' products group, this group also furnishes information and aids the supplier. By an extensive program of national and local advertising in every medium, the firm's name is kept constantly before the buying public. In package advertising, suppliers buy space in trade magazines and journals to promote standard materials and products. Integration of the advertised product image and the image projected by the package is essential to success.

Direct mail campaigns are also important, especially to announce new or special products or to bring attention to specific materials. They can take the form of magazine article reprints and ads, letters, and brochures. Direct mail as a form of advertising is rapidly growing. In 1980 it was $7.6 billion, and by the year 2000 it is projected to reach $27.5 billion.

Fig. 2.7   *An example of excellent photographic reproduction on the package. The outer carton was manufactured by laminating 215 gsm white coated chipboard (printed 4 colors plus varnish) to a double-faced board. Inner fittings are die cut in such a way as to protect and display the contents. Courtesy, Decorprint Division of Duffin Containers, Ltd., Derbyshire, England; graphics by John Russell Packaging, Stoke-on-Trent.*

Fig. 2.8 *Industrial products offered for retail sale in shrink packaging and skin packaging. Both methods guard against pilferage and deliver a fresh package to the consumer.* Courtesy, *Plastics Division, Imperial Chemical Industries, Ltd., Hertfordshire, England.*

In addition to preparing advertising materials for publications, radio, and TV, and direct mail campaigns, advertising personnel also prepare displays for trade shows and conventions. Point-of-sale aids, motion pictures, and other forms of promotional materials are developed in close coordination with the Product and Market groups.

Packaging Research

The packaging research group has three primary goals:

1. To improve the packaging of products and processes
2. To develop technical packaging data for the company and its customers

3. To assist customers in individual packaging problems that arise within their own plants

These functions are performed by a staff of scientists and technicians who are knowledgeable about all forms of materials and have extensive experience in all phases of packaging.

Although packaging is considered to be a support activity by many user firms, it is still not necessarily subservient to the departments whose need it serves. It bears the same relationship in its operation to the positions occupied by lawyers and their clients. Unnecessary costs and ill-feeling in the marketplace are matters that no management can afford to overlook. A well-defined and well-run packaging function often provides an avenue for increased sales and profit to the firm involved.

3   THE PRODUCT IS THE PACKAGE

You hear that the peapod is the perfect package and thus eliminates one middleman. Granted the peapod is attractive and it keeps the product fresh, nicely in place and well protected. But its designer—or should that be a capital "D"—didn't have to list the sodium content or calories per serving or wonder what colors the other peapod designers were using that year.
Casebook Jurors (1980–81)

Package design as we know it today has its roots in the economic hardships of the Depression years. The 1920s were boom years for the print media. Easy dollars bought reams of product advertising in newspapers and magazines. But this lavish spending ended with the Crash, and manufacturers and food processors soon discovered an economical alternative—the advertising potential of the package itself. Although not a 20th-century concept, this was the first time that these ideas had been formalized by groups of independent product manufacturers. The marketing philosophy that "the product is the package" actually began in the 19th century with the introduction of decorated potlids and biscuit tins in mid-Victorian England. Using color on a package to identify its contents can be dated to about the same time.

Fig. 3.1   *Uneeda Bakers (Nabisco) tin, c. 1925, an early example of package design. Courtesy, Landor Associates, San Francisco: Museum of Packaging Antiquities.*

Packages in various colors for home remedies were introduced by the Mason Company in 1898. Yellow pills in a yellow box "cured" dyspepsia, brown pills in a brown box relieved constipation, red pills in a red box suppressed coughs, and white pills in a white box soothed sore throats.

Soon package design became a profession in itself. The research methods pioneered by Walter Dill Scott in advertising were soon applied to package copy, color, shape, and size. The revolution in label design and package styling had begun. Products could sell themselves and compete better on the shelf with strong, motivational labeling. As the self-service era emerged and flourished, even more emphasis was put on attractive, eye-catching designs.

Although package design grew rapidly in the 1930s, its workers actually began their activities in the mid-1920s. Artists such as Gustav Jensen, George Switzer, Walter Dorwin Teague, and Joseph Sinel in the United States, Robert Burns, Reco Capey, William Larkins, Norbert Dutton, Bernard Griffin, Jesse Collins, and Frank Mortimer in Britain, and Egon Schmid and Erich Simon in Germany realized that the intensive marketing of products in branded packages demanded top designs.

One pioneer in design was Lucien Bernhard, a German designer who had recently settled in the United States, and by the early 1930s he had developed a distinctive style. Package design grew rapidly in the next decade, and by 1955 the era of "total marketing" had begun. The package was now considered in terms of its overall advantage to the entire distribution system, particularly to the end of the line—the consumer. Good package design could sell concepts and ingenuity as well as the product inside. The package designer became the liaison between the product and the consumer, with the goal of having the consumer equate product with package. The package had become part of the product, and

Fig. 3.2 *Early spice boxes. Left: tin with paper label, c. 1910. Center: tin with paper label, c. 1930. Right: tin in two colors, c. 1890.* Courtesy, *Collection of Edward Morrill, Werbin & Morrill, Inc., New York City.*

one advertising agency announced candidly, "we couldn't improve the product, so we improved the package."

Package users must either specify a design or obtain the services of a professional package designer. Sometimes the only difference between two competitive brands lies in the packaging. The five factors involved in the integration of packaging into the overall marketing function illustrate the dynamic nature of the packaging process.

1. The package must be able to serve the total retail environment. Packages designed for supermarket sales are different from ones intended for industrial distribution. Packages intended for duty-free markets are also often different than those sold domestically. Vending machine designs differ from convenience-store packages. Shelf life, stacking ability, and impulse appeal must be considered.

2. Package color, shape, and design are also important. Brown bread wrappers and earthy colored cartons convey a natural appeal to the consumer. Shape became a viable factor for the increased sales of Canadian beer in the United States. The bottle used in Canada for LaBatt's beer is short, squat, and brown; however, LaBatt market researchers found out that a premium imported beer image can only be projected by a long slender green bottle. The net result was two different shapes: one short Canadian bottle and one long slender green bottle designed for American distribution. It appears that Heineken's green bottle is the Cadillac of packaging and triggers something in the consumer that identifies it as distinctive from domestic beers.

3. The package must boost sales of the product by the use of special promotional efforts. Should it be a premium pack, bonus size, coupon pack or generic labeled package? These all convey a different image to the consumer and directly affect the pocketbook of the manufacturer.

4. The package must be designed, manufactured, and sold in conjunction with a fully coordinated advertising program. The purpose is to make it instantly recognizable to the consumer.

In a package design survey conducted by R. Overlock Howe and reported in *Advertising Techniques* (September, 1980), eight reasons are given for developing new package designs:

1. Line extension
2. Response to competition
3. New formulation
4. Change in package material
5. Government regulations
6. New products/markets
7. Metric conversion
8. Cost reduction

The reasons for proceeding on the above factors were based on the use of consumer research, in-store competitive evaluation and the use of an outside package design consultant. However, first it was essential to answer the following questions:

1. *Is there a need for a new package?* There should be a definite reason for redesigning the package and a specific need for the redesigned container. When the British firm of Meredith and

Fig. 3.3 *This graceful Champagne Des Princes decanter, intended for the luxury market, was designed for the Prince Imperial who later became Emperor Louis Napoleon III. Photograph reproduced by permission of Maison DeVenoge, owner of the De-Venoge, Champagne Des Princes, Cordon Bleu, and Distinctive Bottle Design trademarks.*

Drew reintroduced their "Crispi Crisps" in new packaging, sales increased by 50 percent. Scores of other successful case histories exist both here and abroad.

2. *What market does the product serve and what is its primary appeal?* Is the product a necessity or a luxury? From January to July 1980 more than 700 new products were introduced. Those products were both food and cosmetic items.

3. *Does the product have a secondary appeal?* Many products serve varied markets, but often these markets become evident only after the product has been introduced. By overlooking the secondary appeal, the advertiser may automatically reduce his chances of widening his market. Demographic trends often play an important factor in developing secondary markets. The introduction of aerosol insecticides in World War II spurred many secondary uses—hairsprays, paints, etc. Convenience and consumer appeal were the main factors in spurring on secondary markets.

Personna Face Guard is an excellent shaving product specifically designed for sensitive skin. Yet, although the product is excellent, an important secondary market has been overlooked for this product. The facial skin of black men tends to be bumpy from ingrown hairs. Yet, Personna Face Guard has not positioned their product against this market.

4. *What is the purpose of the package?* Obviously, the package must sell the product, but it also may have other functions. In a period when many consumers are looking for bargains, the introduction of a large economy package may be quite successful. Coupons can also be integrated into the package as part of the graphics. In 1979 American consumers received about $610 million from manufacturers for redeemed coupons.

5. *What are general design trends?* Package design often closely follows the design trends of the time (See Table 3.1). In checking trends, the food processor must ask himself several questions. What phases of current design are temporary? Which seem likely to last? Will it be profitable to redesign the package to meet temporary changes with the idea that another design can be made as new trends appear?

TABLE 3.1   PACKAGE DESIGN TRENDS: 1975–1985

Trend	Product
Increase in pouch packaging	Retort pouch ("Specialty Sea Foods") Shasta soft drinks "Pockit" fruit drinks
Packaging as an art form	Shiseido cosmetics Stanley garden tools Push Pin Studio's "Puspinoff" Candies
Ecological packaging	"In-A-Box" containers Hasbro's Tente toys Kraft cartons
Black in food packaging	La Crosta pizza mix "O My Goodness!" noodles

6. *What are the design trends in the industry?* The package user who wants to be a design leader may risk failure because of his innovations. If the innovative packaging is expensive, competitors are less likely to copy it.

7. *Who are the prospects and customers?* How and where will the product be used? In what income class are the majority of the prospects and customers to be found? If the market served crosses all income levels, the package user must find universal appeal.

8. *How will the package be displayed?* Will it be used primarily on the counter, in showcases, or on open display tables? By changing the package, the retailer can often be induced to use it more widely on his counters or in his windows.

9. *How will the package be advertised?* The package user must determine whether or not to use the package in the advertisement and then determine the general style and theme of the advertising. How can the package reinforce the main appeal of the advertising? With more than 10,000 individual product names and labels in the supermarket, package naming and advertising becomes critical to success.

10. *What kind of packaging material should be used?* Is the present material satisfactory, or would new materials be more suitable?

11. *Can the package be made profitably?* Will a new design necessitate costly changes in packaging machinery? Can the

design be put on the labels? Will the design be compatible with an outer shipping container?

12. *Is the product part of a family of products?* Some processors favor the introduction of a product line with similar designs to establish corporate image.

13. *What should the product be named?* The naming of the product is of great importance. Many products fail because of a poor name. "Heavenly Hash" is a poor name for an ice cream since it connotes a totally different product. The name "Clackers" may prove to be suitable for a graham cracker based cereal but hardly useful for a package of cigarettes. But to illustrate that no rule of marketing is perfect, consider the success of the J. M. Smucker Company. They turned their name into a super success: "With a name like Smuckers, it has to be good!"

14. *Will consumers like the new package?* Consumer testing is extremely important. Advertisers often spend thousands of dollars in market research, but they fail to test the package design. The importance of proper package design cannot be overemphasized. In a test involving instant coffee, three packages were tried. One specially designed container had passed tests in psychological appeal and display effectiveness. The second was a standard jar. The third was a standard can. Consumers were asked to use the three coffees and report two weeks later on which one they liked best. More than 85 percent wanted the coffee in the specially designed container. When they were told that the same coffee was available in the can or jar, consumers replied that they didn't care so long as it was the same coffee.

GRAPHICS

The package must be able to convey its *message* through graphic design, as well as describe the contents and how to use them. Illustrations or symbolic designs convey direct or indirect messages about the product and its quality and value. The design and copy are important not only to attract attention to the package, but to communicate the desired information. Copy should be simple, legible, complete, and attractive to harmonize with the overall package design.

It is best to keep copy to a minimum and use color, design, and symbols to convey the sales message. The lettering should supplement the package's color plan. Color has a specific effect on legibility. Black lettering on a yellow background is very legible, whereas yellow on white is not. Also, copy is less legible in capital letters than in lowercase ones, and a word is more legible if the space between the letters is larger than the type thickness. The message containing the ingredients and weight should be in legible colors. The purchaser must not think that the label is concealing these facts.

Copy design can convey other messages too. For example, a homegrown-quality image can be conveyed through fancy, curly lettering. To modernize a product, graphics should have straight, simple lines.

COLOR

The entire concept of generic-food packaging is dependent on color. Think of a generic product and a black-and-white package comes to mind. Frugality and economy is implied by the use of two simple basic colors. The consumer is led to believe that simple packaging, devoid of multi-color printing, leads to economy. He or she does not realize that the cost differential is largely due to inferior quality products and off-grade items. However, Loblaw's, a Canadian grocery chain, uses distinctive black-and-yellow packaging for its generic products, believing that black and white labels impart a "feeling of failure," whereas black and yellow signifies "smartness and sophistication."

Color differentiates objects to the eye, independent of their form. Many rules apply to the use of color in package design. Some colors are associated with certain products, and some have psychological associations (see Table 3.2).

Red is a powerful color because it stimulates the digestive system and the circulation of blood, arouses the forces of self-preservation, and signifies strength and virility. European and Chinese doctors used red coloring in the treatment of skin diseases even in the Middle Ages. They believed it brought fresh blood to the skin's surface, facilitating the healing process. In

TABLE 3.2   COLOR EXPRESSION CHART

Product	Color Identification
Milk	Dark blue/pale blue/white
Coffee	Brown/gold
Spices	Green/gray; red when they are strong (paprika)
Chocolate	Red/orange (sweetness of chocolate), brown/pale blue
Frozen foods	Bluish green/white
Detergents	Blue/white, eventually green
Razor blades	Steel gray/medium blue
Menthol cigarettes	Pale green/white
Insecticides	Yellow/black
Baby products	Pale pink/pale blue
Perfumes	Violet/lilac
Furniture polish	Brown shades

SOURCE: "Color Sells Your Package," Favre, *ABC Edition Zurich* (1969). These colors are based on a series of consumer interview tests and are combinations which facilitate product identification.

most cases, the use of red must be carefully controlled. Light red is a cheerful color, but dark and bright red are more likely to induce depression and irritation. Cherry red is sensuous.

Orange, more subtle than red, is often used on packages associated with the physical because it expresses action. It looks clean and appetizing and has the cozy, intimate character of a fire burning in a fireplace.

Yellow denotes, light, gaiety, and warmth, and it is cheerful and bright. It is also a preferable color for use in the Far East. Pale yellow looks dainty, golden yellow is active, green-yellow is sickly, and a deep, strong yellow suggests sensuousness.

Pink suggests femininity and deep affection. It lacks vitality and gives an impression of intimacy and gentility. A bright magenta pink is associated with frivolity.

Green, quiet and refreshing, is associated with youth, growth, and hope. It is undemanding, evoking neither passion nor sadness, although in Arab countries, it is a sacred color, the symbol of the prophets and Islam, and should never be used in package design. When darkened to olive, it becomes a symbol of decay.

Blue is cool and subdued. Whereas green suggests earthlike quiet, blue suggests heavenlike quiet.

Color can be used as an attention getter as well as appealing to the emotions. The best attention-getting colors for males are black, red, orange, green, then yellow; for females, they are red, green, black, orange, then yellow. Attention value in color is also determined by the size of the color areas. Certain color combinations will win attention for a small package but may seem garish on a larger one.

One interesting example of a planned program of subtle communication is a Swiss milk package. Before the "Tetra-Pak" unit was introduced in Switzerland, in competition with glass bottles, the new package was carefully evaluated by Swiss consumers. The suggestions for the choice of color and the survey are derived from the Central Association of Swiss Milk Producers. With the help of Scope Ltd., Market Research Institute, Lucerne, Switzerland, the association carried out a survey on the Swiss market in order to get to know the habits of consumption and the general attitude toward milk.

Color was of prime importance in consumer communication. Tests revealed that, in order to denote cleanliness and purity, light colors, a white background, and modern design motifs should be used. Consumers associated red, orange, and brown with fat. To play this down and suppress the idea that milk is fattening and high in cholesterol, these colors were avoided in favor of light colors. To communicate that milk is a thirst quencher, various shades of blue with fresh, light colors were found to be best. To communicate that milk does not have an unpleasant taste, vivid color contrasts were found to be best. To imply good value, sophisticated design and multicolor printing detail were employed. Similarly, to overcome the feeling of many men that milk drinking is effeminate and childish, strong masculine and vigorous colors were used with linear and angular designs. Red, considered a masculine color, also appeals to a broad range of socioeconomic levels. The net result was a modern-looking package with blue, white and red.

In a highly competitive category like shampoos, Vidal Sassoon depended on packaging to distinguish his from the pack. The brown bottle is understated in a segment where bright color usually sells. The package created the character.

Minnetonka's "Softsoap" package designers conceived a strategy related to product form, product dispensing, and packaging, and they spent a considerable sum of money on package design. The product has been a notable success.

Louis Cheskin, director of the Color Research Institute, has done much of the pioneering work in the use of color in package design. In his many books, such as *Color for Profit, Color Guide for Marketing Media, and How to Predict What People Will Buy,* he often stresses the role of a color in determining both the optical and psychological effects of a package.

DISPLAY UNITS

The sales value of a package usually depends to a great extent on how the unit is displayed. Even the most attractive container can do only a mediocre job if it is put high on a shelf in the back of a store. Because the package does its most important work at the point of sale, display is a crucial factor in package design.

The four types of display to consider are (1) shelf, (2) counter, (3) window, and (4) floor.

Shelf

Will the package fit onto the shelf easily? One of the inherent disadvantages with the tetrahedral package is its unique shape, which makes shelf stacking difficult. In addition, consumers can easily disrupt a stack of packages and leave an unsightly display. Shelves have become standardized in most retail outlets, with the result that packages of certain heights may be stored behind counters if they do not fit on standard-size shelves.

Counter

It is on the counter that the package does its most important sales job, because it is there that the package first comes into contact with the customer. Often, specially designed outer cartons are used to enhance counter appeal. This is particularly true for small packaged items, such as confections or batteries. These cartons are placed on counters and consumers are encouraged at the point of purchase.

Fig. 3.4 *This superb package design is aimed at making the consumer more conscious of the name Dixie as a sign of quality and dependability. Courtesy, Irv Koons Associates, Inc., New York City.*

Visual Displays

The product must be visually attractive. An effective visual display can do wonders in selling a product. Canned goods suffer from the fact that their contents cannot be seen. The entire selling job depends on the label and often the vignette on the label represents the can's contents poorly. Properly designed visual displays can solve this problem. One way to handle this is to pack an actual visible sample of the product in a little glass jar. This is placed on top of the cans in the display. Backed up by a display card, an excellent visual display is obtained.

Floor Displays

In open display, the general effect should be asymmetrical. Retailers found some years ago that the average person dislikes selecting a package from a symmetrical display because he hesitates to break it up. The packages were either jumbled together

Fig. 3.5 *A Dundee wooden chocolate box designed as a treasure chest. Chocolates shaped as orange slices are wrapped in gold aluminum foil, bundled in groups of 20, then wrapped in printed aluminum foil to form a sphere. "Oranges" are then wrapped in orange cellophane with green leaf decor and packed in a wooden box with an abaca rope and overwrapped with transparent polypropylene. Courtesy, Goya Products, Inc., Marikina, Philippines.*

or, where stacked, avoided neat, pyramid or cube effects. It has also been found that the most effective floor display is arranged so that the consumer gets the impression that the original arrangement was symmetrical, but that customers have already bought a number of packages.

DESIGNING PACKAGES FOR EXPORT

Packaging for export is highly specialized. The American package user should never put his package directly into foreign circulation without first consulting retailing authorities in the countries to which he will export. Some packages should never be changed, such as a French perfume bottle and the label on a Scotch whiskey bottle. A French association with perfume or a Scotch association with blended spirits is a strong selling point.

Many times, developing nations will use the printed package to convey a certain message. U.S. manufacturers exporting protein-rich infant food packages to Pakistan have used printed polyethylene bags that carry a birth control promotion message.

A package such as this is obviously unsuitable for domestic sale but highly desirable in the Third World.

In formulating a package for export sale, the American package user should consider the following factors:

Climate
Language (make sure directions for use are easily understood by inhabitants of the nation involved)
Literacy
Trade/Living Customs
Shipping Conditions
Product use
Inserts (all instructions should be written by someone thoroughly familiar with the language)

An additional point: Different nationalities perceive different colors in different ways (see Table 3.3). In the United States, red denotes cleanliness, but in Great Britain, it is the least clean of

Fig. 3.6 *The Monkey symbol on this Guinness Stout export label indicates the probable market as India. Courtesy, Guinness Museum, Dublin, Ireland.*

all colors. In Sweden, blue is masculine, but in Holland, it is feminine. Red is preferred in Italy, whereas Great Britain, Sweden, and Holland prefer blue and golden yellow.

Colors that evoke an immediate reaction in an American consumer often do little to the Spanish consumer. Green means cool or minty to the American but nothing special to the Spanish, and black is used frequently in food packaging in Spain but rarely in the United States, although there are two notable exceptions. La Costa Pizza Mix and O My Goodness! Noodles successfully used black or dark brown fading into black on their packages. Black, especially in film pouches, protects the product from light.

The importance of symbolism as a factor in package design can be illustrated by the case of an exporter who planned to use a reproduction of the Venus de Milo on a package that was to be exported to an Arab country. Then he discovered not only that the traditional punishment for criminals in these nations is to have his hands chopped off, but that the Koran prohibits the reproduction of any living creation.

TABLE 3.3   COLORS AND SYMBOLS FOR PACKAGED EXPORTS

Nation	Use	Avoid
Ethiopia		Black
Nigeria		Pairs
Sudan	Green	Black, red, yellow (sickness)
Tunisia		Black
Republic of South Africa		Black
Afghanistan		Black
Burma	Gold, yellow	Purple
Ceylon	Gold, yellow	Purple, black (Representation of Buddha or representation of the be-tree, a religious symbol)
China and Taiwan	Yellow (Tigers, lions, dragons (strength), elephants	White, black, blue (mourning)
Hong Kong	Yellow Tigers, lions dragons (strength)	Black, blue
India	Green, orange, yellow Monkey	Black Cows, dogs

Nation	Use	Avoid
Indonesia	Green	Red, white, blue (reminiscent of colonial rule)
Iran	Green	Gold, yellow, black
Iraq	Green	Yellow
Japan	Gold, silver, white purple, yellow, cherry blossom	Pink, red, black (The number 4, which in Japanese is pronounced like the word for death)
Jordan	Green	Dark violet
Korea	Yellow	Olive drab, white
Laos	Gold, yellow	Purple
Lebanon		Silver
Malaysia	Gold	Yellow, green Cows, pigs
Pakistan	Green, orange	Violet, black (Representation of Mohammed, representation of pig or monkey)
Philippines		Violet, black
Saudi Arabia		Black
Singapore	Red, red and gold, red and white	Snakes, pigs, cows, tortoises
Thailand		White, black (mourning) (Representation of feet regarded as despicable)
Tahiti	Red, green, gold, silver	Elephants
Vietnam		Black (treachery)
Yemen		(Representation of man or beast)
Australia		Black The red rising sun (Japan's colors World War II)
New Zealand		White, red, or silver crest
Austria		Black
Belgium		Violet, black
Cyprus		Red
Czechoslovakia		Gray green
Denmark		Black, yellow
England	Blue, yellow	Purple
France		Black, green
Germany (East/West)		Black, brown (in shirts)

(Table continues)

TABLE 3.3—Continued

Nation	Use	Avoid
Greece		Black
Hungary		Black, yellow and black Cross and Arrow
Italy	Red	Black
Lithuania		Black
Switzerland		White cross on red Red Cross
Yugoslavia	Red	Black
Costa Rica		Gold, red, white, .blue
Cuba		Black
Guatemala	Light colors	Red, gold
Haiti	Red, light colors	Violet, black green (in shirts)
Mexico	Red, light colors	Green, white, red
Nicaragua	Red, light colors	Violet, black
Argentina	Pastels, light colors	Black, light brown
Chile	Pastels, light colors	Bright red
Colombia	Light colors	Black, red
Venezuela	Light colors	Gray, yellow, blue, red

SOURCES: Hygrade Packaging Company (New Zealand), and Charles Winick, "Taboo and Disapproved Colors and Symbols in Various Foreign Countries," *J. Sec. Psych.* 59, (1963): 361–68.

4 THE PACKAGE AND THE PROMISE: THE ROLE OF ADVERTISING

The man who doesn't believe in advertising is constantly doing what he deprecates. He hangs coats outside of the door, or puts dry goods in his windows—that's advertising. He has printed cards lying on his counter—that's advertising. He sends out drummers through the country, or puts his name on his wagon—that's advertising. He labels the articles of his manufacture—that's advertising. A man can't do business without advertising, and the question is whether to call to his aid the engine of the world—the printing press, with its thousands of messengers working night and day; the steam engine adding to its replete capacity and untold power and miraculous speed: or, rejecting all these, to go back to the days when newspapers, telegraphs, and railroads were unknown. "But advertising costs money!" So does everything that is worth having. If advertising cost nothing, all the third, fourth, and second class petty shops would stand an equal chance with the most respectable houses. If you want to prove to the world that yours is a first class establishment, advertise.

Daniel Frohman, *Hints to Advertisers* (New York, 1869)

How to advertise and sell the package and product is often the most difficult task for the producer. He usually lacks both the sales and marketing knowledge needed for the success of his products. This is why the vast majority of new products fail. In a recent survey conducted by the A.C. Nielson Company on fast-turnover consumer products, the interaction between the product-package was responsible for about 53 percent of the failures of unsuccessful new products. Within this category there were scattered instances of the product being too far ahead of the times or of the wrong package size. Advertising quality, media, and appeal (including product name) accounted for another 10 percent of these failures. More than 60 percent of all new products fail because of these factors.

THE EARLY YEARS OF ADVERTISING

It all started with the pioneering articles of Walter Dill Scott, director of the Psychological Laboratory of Northwestern University. His "Psychology of Advertising" series was published in

Fig. 4.1 *This classic, French turn-of-the-century Art Nouveau poster by Privat-Livemont extolled the virtues of Rajah tea and coffee. Courtesy, Phillips, The International Fine Art Auctioneers, London, England.*

the *Atlantic Monthly* in 1903. Then a professor of psychology at Northwestern, Scott pointed out the defects of advertising copy and stressed the need for cheerful advertisements. He criticized those advertisers who linked their products, even in jest, with bulls, pigs, frogs, and other creatures. It was also a mistake to associate the product with undesirable characters. In his articles he shifted the focus from explaining how a product worked to describing the pleasure it would give the user. In short order, the advertising profession began to adopt the terminology of the psychologist.

Around 1910 departments were started in many advertising agencies to gather nationwide market data. Advertising plans based on these data resulted in copy derived from detailed knowledge of the product, its market, and its uses. Slogans were soon adopted and advertising agencies began to prosper.

The print media boomed after World War I and the postwar recession, and advertising expenditures have steadily increased since, with more than $300 billion projected for the year 2000 (see Tables 4.1 and 4.2).

THE ADVERTISING AGENCY

The number of advertising agencies has increased annually, and today there are more than 6,000 in the United States. Personnel have also increased. From 1970 to 1979, the J. Walter Thompson Agency has increased from 2,830 employees to 6,200 employees.

Most advertising agencies operate on a retainer basis and offer a wide variety of skills to the manufacturer. A typical agency

TABLE 4.1 ADVERTISING EXPENDITURES (1940–2000)

Year	Amount (in billions)
1940	$ 2
1950	$ 5.7
1960	$ 11.3
1970	$ 19.6
1979	$ 50*
1980	$ 56.8
1990 (projected)	$140.5
2000 (projected)	$305

*The top 100 advertisers spent $11.7 billion in 1979.

TABLE 4.2   100 LEADING NATIONAL ADVERTISERS
(TOTAL AD DOLLARS IN MILLIONS, 1979;
P = COMPANY MARKETS PACKAGED PRODUCTS)

1. Procter & Gamble	P	$614.9
2. General Foods	P	393.0
3. Sears, Roebuck		379.3
4. General Motors		323.4
5. Philip Morris	P	291.2
6. K Mart		287.1
7. R.J. Reynolds	P	258.1
8. Warner-Lambert	P	220.2
9. American Telephone & Telegraph		219.8
10. Ford Motor		215.0
11. PepsiCo	P	212.0
12. Bristol-Myers	P	210.6
13. American Home Products	P	206.0
14. McDonald's	P	202.8
15. Gulf + Western Industries		191.5
16. General Mills	P	190.7
17. Esmark		170.5
18. Coca-Cola	P	169.3
19. Seagram's	P	168.0
20. Mobil		165.8
21. Norton Simon	P	163.2
22. Anheuser-Busch	P	160.5
23. Unilever U.S.		160.0
24. RCA		158.6
25. Johnson & Johnson	P	157.7
26. Heublein	P	155.0
27. Beatrice Foods	P	150.0
28. CBS		146.1
29. U.S. Government		146.1
30. Loews		144.5
31. General Electric		139.4
32. International Telephone & Telegraph		132.4
33. Pillsbury	P	131.5
34. American Cyanamid		127.0
35. Gillette	P	126.9
36. Richardson-Merrell	P	123.8
37. Colgate-Palmolive	P	122.5
38. J.C. Penney		122.0
39. Kraft	P	119.7
40. Chrysler		118.0
41. B.A.T. Industries		116.4
42. Ralston Purina	P	108.0
43. SmithKline	P	107.7
44. Chesebrough-Pond's	P	107.3
45. Consolidated Foods	P	105.0
46. Time		102.4
47. Revlon	P	101.0
48. Transamerica		95.0
49. Sterling Drug	P	92.0

50. Kellogg	P	91.6
51. Nabisco	P	91.3
52. DuPont	P	89.4
53. Eastman Kodak	P	87.8
54. Quaker Oats	P	86.6
55. Nestle Enterprises	P	86.0
56. American Brands	P	83.5
57. Toyota Motor Sales U.S.A.		80.3
58. Schering-Plough	P	78.0
59. Miles Laboratories	P	77.8
60. Clorox	P	72.6
61. CPC International	P	72.0
62. Jos. Schlitz Brewing	P	71.6
63. H.J. Heinz	P	71.5
64. Mars	P	69.5
65. Nissan Motor		66.1
66. MCA		66.0
67. Mattel		66.0
68. Trans World		62.0
69. Campbell Soup	P	60.5
70. Squibb	P	60.0
71. Liggett Group	P	59.0
72. Warner Communications		57.6
73. American Express		55.4
74. Union Carbide		55.1
75. Volkswagen of America		55.0
76. Greyhound		53.8
77. UAL		52.5
78. Polaroid	P	50.5
79. Brown-Forman Distillers	P	50.0
80. MortonNorwich	P	49.2
81. Wm. Wrigley Jr.	P	48.5
82. Beecham Group	P	46.7
83. American Motors		44.6
84. North American Philips		44.2
85. American Honda Motor		44.0
86. Pfizer	P	43.7
87. ABC		42.0
88. Eastern Air Lines		40.4
89. Noxell	P	40.2
90. S.C. Johnson & Son	P	40.2
91. Borden	P	39.0
92. Levi Strauss		38.6
93. A.H. Robins	P	37.0
94. Scott Paper	P	36.8
95. Standard Brands	P	36.4
96. American Airlines		35.0
97. Delta Air Lines		33.5
98. Milton Bradley	P	31.3
99. International Business Machines		31.1
100. Mazda Motors of America		28.4

SOURCE: *Advertising Age* (September 11, 1980).

staff includes package designers, marketing experts, and perhaps even a packaging technologist. Although the latter is still a rarity, several large agencies either employ or have as consultants topnotch package-development specialists.

Before choosing an advertising agency, the wise product user will do the following:

1. Assess agency personnel
2. Assess agency philosophy
3. Examine several of the agency's past campaigns
4. Determine whether the agency provides short-lived themes or central concepts that last several seasons

An advertising agency can be expensive, so the product user should select carefully the jobs he passes on to it.

The agency should work closely with the designer and the packaging director in making packaging suggestions at the outset of the advertising campaign. It should also coordinate the design program with the advertising campaign and make certain that both promote the same selling message. The agency should be allowed to offer its suggestions on writing package copy. All too often this is done piecemeal and as an afterthought by the manufacturer. Copy writing is a skill, and the talent of a copywriter often means the difference between success and failure. Good copywriting often results from a two-way communication between the writer and the art director of the advertising agency. There are no absolute rules on how to write good, saleable copy, but a good copywriter knows the product, the campaign, and the needs of the customer.

The role of the advertising agency often changes over the years. While the 1970s can be referred to as the Era of Product Positioning (how it was said), the 1980s is the Era of Consumer Positioning (what is said and to whom). The role of the consumer has become more important with increased market segmentation.

THE PACKAGE IN ADVERTISING

Whether or not a firm uses the talents of the advertising agency, there is still a definite need to create an *identity* for the product.

The success of an advertising campaign relies on a sound marketing objective: Consumers should request the product in the retail store. Creating the *package profile* is not an easy task. There are several simple rules to follow, but they are not foolproof:

Show the package in the advertising. The possibilities of using the package in all types of advertising are virtually limitless. It is more effective to show the package in use rather than by itself. An action illustration is more effective than a picture of the package alone. A consumer should be shown drinking a bottled beer or using an electric drill. This "use" connotation is a more effective sales tool for the product.

Fig. 4.2 *Pathmark's line of generic products.* Courtesy, *Lippincott & Margulies, Inc., New York City.*

Show the package opened. An open package conveys an effect of liveliness. This rule cannot be applied universally though; for example, an opened can is not particularly inviting, but a well-designed can unopened has decorative value.

Use color. The use of color in advertising offers many obvious benefits to the food manufacturer. The chief reason is to establish identity and enhance recognition that will carry into the store or supermarket. Color makes a food more appetizing and can convey certain moods to the consumer.

Design advertisements carefully. Never approve an ad that is not realistic simply to highlight the package. A butler holding an open can of vegetables for the admiring gaze of a roomful of guests in evening gowns is hardly realistic. Such an advertisement was actually published in a leading magazine.

To advertise the product to its best advantage is no mean trick, but several standard methods apply:

1. Feature the package as the main point of the advertisement. The "Marlboro Man" is a strong personality, but ads focus in on the cigarette pack and not on the user.
2. Make the package an accessory feature to the main illustration. A woman using a cosmetic or a man offering another a cigarette are illustrative subjects where the package is an accessory.
3. Use the package merely to establish identity. The advertisement copy and illustrations may have no reference to the package, but serve to fix the design in the consumer's mind.
4. The sides of the package should not be parallel to the ad's margins. Greater effectiveness is usually achieved if the package is tilted.

The package is part of a general marketing plan and, of course, how it is used in that plan depends not on the package, but on the basic thrust of the marketing strategy.

THE MEDIA

More than $200 per capita per season is spent on advertising in the United States (compared with only about $80 per capita in

France), so selecting the right media for advertising obviously is extremely important. Television, often called advertising's "third dimension," is the most important advertising medium. It tells us what to buy and where to buy it. A full-color package that is effective on the retailer's shelf may have little or no impact on black-and-white TV (although it will probably be acceptable on a color set). Television advertising must be able to project the brand name and package identity to the consumer.

Some of the most popular campaigns succeed in promoting the use of a whole product group at the same time. Sony's promotion of videotapes serves to educate the public as to the product and even helps Panasonic sell their brand of tapes. One way to eliminate this problem is for the Sony package to have clear product and manufacturer identification. This can minimize the positive effect a Sony campaign might have on Panasonic products. This is true of the name of the product as well as the specific design. When the Reynolds Metals Company advertises Reynolds Wrap, the use of household foil in general tends to increase. Many consumers think of Reynolds Wrap as generic. Here is a case where a trade name has clearly been identified with a specific product line. Additional examples include Kleenex as a generic term for facial tissues and Saran Wrap as generic for plastic household wrap.

Television combines visual and audio impact, but radio has audio impact only. Television, radio, and the national or local press are ideal for advertising consumer goods, but industrial goods are best advertised in the trade or technical press. Occasionally, industrial advertising in the prestige national dailies can be valuable because they are widely read by top management. More than $74.2 billion was spent on newspaper ads in 1980. *The Wall Street Journal* ran several advertisements in 1979 for the retort pouch (a flexible, shelf-stable food pouch) and its uses. This was an excellent example of an ad intended to reach a wide cross section of executives and inform them of a novel package ready to hit the market.

Packaging or advertising changes, of course, are not made independently; they are made by considering the effect one may have on the other. A simple change in color, for instance, may make an item unrecognizable on television.

THE DUAL-USE PACKAGE

One of the most effective methods of advertising a product and keeping it in the public eye is to design and produce a package that has multiple uses. There are many examples of dual-use packages. One of the earliest was the one-pound peanut-butter can that could be used by children as a sandbox toy after the product was consumed. Other examples include a mustard jar that later becomes a beer stein, a bean pot that holds smoking tobacco, or cheese packages that become drinking glasses.

There are two types of dual-use packages. One variety has no relation between primary and secondary use. One example is the liquor bottle that later becomes a table lamp. In this case, only if one retains some type of product identification can the product remain in the consumer's mind. The second category has a close-ly related use. One example is the foil tray that holds cake mix and becomes a reusable cake pan.

The dual-use package has many important advantages for a food manufacturer. It increases sales appeal at time of purchase, and it serves to remind the user of the product long after it has been consumed. The more ingenious dual-use containers also have a distinct novelty value. But dual-use containers also have some concrete disadvantages. Psychologically, the average consumer hesitates to discard a dual-use container, particularly since he or she knows it increased the cost of the product. If there is no obvious use for it however, the consumer may turn to another product that offers better value for the money. Therefore, the best dual-use container is the one that bears no extra cost and is just as convenient as an ordinary package.

CHILD APPEAL IN PACKAGING

While an ad campaign can and often does appeal to children, a properly designed package can do wonders as an aide to advertising efforts.

An excellent way to appeal to children is to offer prizes in the packages. Borden's Cracker Jack has held to a policy of packing some simple gift for children in every package.

Fig. 4.3 *Even in the 1890s, advertisements featured actual package images. This one shows a delivery boy carrying out an assortment of Fry's cocoa and chocolates from Harrod's Department Store in London. Courtesy, Sotheby's Belgravia, London, England.*

An unusual package design also can serve to stimulate child appeal. Years ago, General Foods used cutouts of soldiers, animals, etc., on the Post Toasties package. These cutouts became toys for children without any investment on the part of parents.

Packages can also turn into dollhouses and metal cans into seed containers and planters. An early example is Barnum's Animal Crackers. The paperboard carton containing these cookies is designed like a circus cage. Coupled with the animal-shaped cookies, the entire unit is a natural for child appeal. More recently, an Austrian firm introduced a peppermint candy packaged in a dispenser, the head of which was in the shape of different animals. It proved to be an outstanding success.

5 FROM GUTENBURG TO GRAVURE: PRINTING TECHNIQUES

A package must literally shout attention to the product. Yet, after attracting the consumer, it must fade into the background and permit the product to come forward.

Ernest Dichter, *Packaging: The Sixth Sense?* (Cahners, 1975).

Packages are printed by a wide variety of methods. While the methods are well established and their techniques have been known for many years, recent business demands have imposed stringent requirements on how to *print* the package.

In the 1980s the cost of materials will increase, whereas the availability of materials will decrease. In addition, capital investment appears to be rapidly declining and, therefore, the development of new-plant capacity is decreasing. To counteract these trends, new demands will be made on the package designer to come up with more sophisticated packaging. In other words, more has to be done for the product manufacturer for less. The goal in the 1980s will be to keep costs low while maintaining the superior protection of the product.

Good graphics are expensive. Design and print cost money. And, even if money is spent on good design, it must be properly executed by the printer. The most appealing design loses its value if it is improperly or poorly printed. Poor graphics, in fact, are worse than none at all. The product user in today's market will have to have a complete understanding of both the design factors involved in packaging and the mechanical methods required to reproduce the designs. The graphic design on a package must convey both product information and product promotion to the consumer. Because the 1980s will be a "tight decade," a sound knowledge of printing methods becomes doubly important.

THE ORIGINS OF PRINTING

Some historians claim that movable type was invented in the Orient by Pi Sheng 400 years before the experiments of Johann Gutenburg and the printing of the Gutenburg Bible in 1455. Setting type with platen and flatbed presses was relatively simple if the printer had infinite patience, a thorough grasp of lexicography, and a mastery of spelling.

The Fourdrinier machine, invented in 1810 by the English papermakers Henry and Sealy Fourdrinier, and the manufacture of paper in rolls allowed the printing of a continuous pattern. In the mid-1800s, the rotary press, which used rolls of paper was developed. It was a distinct improvement over its predecessors in all but one respect: The movable type, set into a curved frame, worked loose and periodically bombarded the printer with a spray of flying metal. But before long a method for casting curved metal plates from the original type was developed.

LETTERPRESS (RAISED PRINTING SURFACE)

The letterpress process, one of the earliest known printing techniques, employed hand-cut plates before the 15th century. Gutenburg conceived the idea of interchangeable printing elements, one for each letter, which could be assembled in any order to suit

TABLE 5.1   SELECTED PACKAGING/PRINTING DATES

1660	First known printed wrapper for a packaged product: Buckworth's Cough Lozenges, England. There is some evidence that the first wrapper was printed by Andreas Bernhardt, a German papermaker, in the 1550s.
1798	Development of lithography by Alois Senefelder.
1800	Iron printing press: Designed by the third Earl of Stanhope, manufactured by Robert Walker of Vine Street, London; first model installed at Shakespere Press, St. James, London.
1811	Power printing press: Steam-driven Koenig press used by Thomas Bensley to print 3,000 sheets of *Annual Register,* London, April 1811.
1845	Rotary printing press: Hoe Rotary adopted by *Philadelphia Ledger.*
1853	Printing on tin: Litho presses patented by Charles Adams, London, Sept. 13, 1853.
1875	Offset-litho: Developed by Robert Barclay, London, for printing on tin: first used on Bryant and Mary tin matchboxes.
1886	First photographic package design recorded: Signor Valli Cigarettes, Denmark.
1907	Silkscreen process developed by Samuel Simon, Manchester, England.
1914	The New York *Times* first used gravure to print its Sunday supplements.
1921	Flexography first used in the United States to print kraft bags.
1932	Gravure first used for packaging application; an integral in-line operation which included creasing, gluing and folding of cartons.
1960s	Development of electrostatic printing, the transfer of dry ink particles without pressure or actual contact of the printing plate with material to be printed.

the text and the particular job. The printing elements were assembled and clamped together as a "forme" on the flat bed of the press. The ink was applied to the printing areas and the paper pressed upon the inked surface.

Today, three types of letterpress are used in the packaging industry, and each one offers good print quality: platen presses, flat-bed cylinder presses, and rotary presses.

Platen presses operate by conveying both the material to be printed and the type form on two flat surfaces, the platen and the bed, that open and close somewhat like the jaws of a clamshell. The bed holds the type form; the platen holds the material. As the jaws of the press open, the type form is inked and a sheet of the material is fed to the platen. As the jaws close, the sheet is printed. When they open again, the printed sheet is delivered and a new sheet is fed to the platen. On most presses the amount of impression, or squeeze, is controlled by an impression lever.

The advantage of using letterpress is that it is well-suited for small runs and small, simple forms. The platen press can also be

Fig. 5.1 *Sheet-fed labels being printed by letterpress.* Courtesy, General Foods Corp., White Plains; photograph by Drennan.

used for embossing, die cutting, and carton scoring and creasing. The disadvantages are that only cut sheets can be used and only one color can be printed at a time.

Flat-bed cylinder presses generally have a moving flat bed that holds the form while a fixed rotating impression cylinder provides the pressure. The material, held securely to the cylinder by a set of steel clamps known as grippers, is rolled over the form as the bed passes under the cylinder. As the bed returns to its original position, the cylinder is raised, the form re-inked, and the printed sheet delivered. On vertical presses, both the form and the cylinder move up and down in a reciprocating motion. This cuts the usual two-revolution flat-bed motion in half; the impression cylinder makes only one revolution for every printed impression.

Only sheets can be printed by the flat-bed method, and the same problems that exist for platen presses also apply. Two colors can be printed on one side, and flat-bed work can range from simple black-and-white designs to high-quality photographic reproductions. The method is applicable for medium-size runs.

Rotary presses, with both a cylindrical plate carrier and a cylindrical impression member, are by far the most efficient of the letterpress machines. The long set-up time makes them suitable only for long runs, but up to five colors can be printed at one time. The rotary press can print either sheets or rolls. The sheets are fed between the two cylinders, using a plate curved to fit the cylinder.

Letterpress printing is used to print labels, bags, folding cartons, plastic films, and corrugated cartons. Because changes can be made simply, it is an inexpensive process. Coated boards and papers often yield the best results with letterpress. The pasty inks that are used are difficult to apply to most plastic films other than rigid PVC and cellophane, which are fairly stiff and can withstand high pressures.

FLEXOGRAPHY (RAISED PRINTING SURFACE)

Widely used for the printing of all types of plastic films, flexography, an offshoot of letterpress, is a high-speed technique. Flexographic printing originated about 1890, when the London paper-bag maker Bibby, Baron and Sons, Ltd. applied for a patent on a method of rotary printing from rubber plates. The inks used were basically glucose and pigment in water. This was a radical departure from conventional letterpress inks, which consisted of pigments ground in boiled linseed oils. Printers of the day called this new process "Bibby's Folly" since it was a dismal failure. The first true flexographic press is credited to a machinist named C. H. Howley, who developed it in 1905 for printing paper bags. Flexography was introduced in the United States in the early 1920s and was used almost exclusively in conjunction with bag-making machines. When plastic films were developed, flexography became widely used in the packaging industry.

In flexography, a thin, fast-drying ink is applied to the film surface with a flexible rubber plate (or stereotype), mounted on a plate cylinder containing raised characters. Ink is transferred to the rubber plate from the ink fountain via a rubber inking roller and an anilox roller. (The anilox roller is made of engraved stainless steel and holds ink in the recesses of its design, thus acting as a metering device to the flexible rubber plate.) The process offers high-speed printing at fairly low costs. Multicolored printing (up to six colors) at one pass is possible using several printing heads on a single-impression cylinder. A significant disadvantage of flexography is that it is difficult to reproduce sharp images and fine detail. Fine type may fill in, so it is limited to screen rulings of 65 to 85 lines per inch. Finer screens require much slower press speeds to prevent filling in of the spaces between halftone dots.

GRAVURE (DEPRESSED PRINTING SURFACE)

Gravure printing was developed in the late 18th century for printing wallpaper. Often considered to be an outgrowth of the ancient art of etching, the process has always been a rotary one, using either a cylinder whose surface is recessed to produce the printing areas or a thin copper plate etched and then wrapped around a cylinder. These engraved copper intaglio plates, the forerunner of gravure, were first used in France and Italy about 1476. Copper engravings offered competition to woodcuts in England and France in the mid-16th century. Copperplate work is still used for invitations and announcements. Web-fed gravure

Fig. 5.2 *The Waviblock system is a preform and fill system consisting of a roll of presealed sacks. With this system, it is possible to fill automatically: sacks are taken one by one from the roll, filled, then sealed and thrown on a conveyor belt. One roll holds 1200–1500 sacks. One man could operate as many as three Waviblock systems simultaneously. Courtesy, Wavin PFP, Plastic Film Products, The Netherlands.*

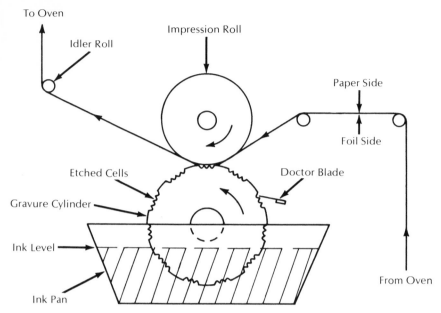

To Oven

Idler Roll

Impression Roll

Paper Side

Foil Side

Etched Cells

Doctor Blade

Gravure Cylinder

Ink Level

From Oven

Ink Pan

Fig. 5.3 *Rotogravure printing station.*

was introduced in the United States in the early 1900s. By 1932 it was being used for printing cartons.

Rotogravure (or gravure) now consists of inking an engraved printing roller that transfers the ink directly onto the material to be printed. The design is made up of a series of tiny cells of varying depth so that different amounts of ink are picked up by different parts of the roller. (The layer of ink produced by gravure on carton stock is about .3mm. thick.) Rotogravure is basically a continuous-tone process. Only in the light and some medium tones are the cell walls visible after printing, and then only under magnification.

Multicolor work must be dried between each color application. After passing through a hot air blast, the film moves over a water-cooled roller to prevent transference of heat from the drying chamber to the next printing unit, where an increase in temperature would cause the volatile ink to evaporate partially and dry in the cells of the engraved cylinder before contacting the web. The cooling roller also hardens the ink from the previous station that was rendered tacky by the application of heat sufficient to melt the resin content of the ink.

Disadvantages of rotogravure printing include the high initial costs of the etched metal rolls and the printing speeds, which are somewhat slower than those obtainable with flexographic processes. The main advantage is that it produces high-quality, multicolor, fine-detail printing.

The high initial cost of the etched cylinders often is a deterrent to small manufacturers of packaged products. The costs involved in purchasing cylinders for a five- or six-color print job may run as high as $5,000. These cylinders remain at the supplier's plant but are the property of the manufacturer. However, when the manufacturer changes packaging suppliers, he must purchase new cylinders. Rarely, if ever, are cylinders sent from one supplier to another. The complexities of individual presses and personnel make this quite difficult. Many product manufacturers believe that this is an excessive charge; but if there is an interesting product to be marketed and it has an outstanding potential volume, a package printer may often absorb the cylinder charge.

OFFSET LITHOGRAPHY (SAME-PLANE TECHNIQUE)

Early lithography was a slow and highly crafted profession. The process required heavy stones, careful alignment, treated paper, and meticulous attention to detail. Even the most basic designs were difficult to lithograph. The inventor of the process, Alois Senefelder discovered near the end of the 18th century that a certain type of porous stone would pick up a greasy ink when dry, but if the surface of the stone was damp it would no longer accept the ink. Early litho machines operated on the flat-bed principle and were as slow as letterpress flatbeds. The stones reciprocated under damping rollers, inking rollers, and an impression cylinder.

The offset principle was devised by tinplate printers, who found that their stones were damaged by the rough edges of the tinplate sheets, making it difficult to get even contact between the stone and the tinplate. They overcame this by transferring the ink image on the stone onto a rubber blanket wrapped around a cylinder; then this image was retransferred to the tinplate sheet by passing the sheet between the blanket cylinder and an impression cylinder. Although offset, this was still a flat-bed process and a slow one. When it was discovered that certain metal surfaces could be treated to be either hydropholic or hydrophilic, flat-bed techniques were replaced by the faster technique of rotary offset. The latter process uses a flexible metal plate on one cylinder, a rubber blanket on the second cylinder, and an impression cylinder with a suitable resilient surface.

In addition to a metal printing plate and an impression cylinder, offset presses now use an offset cylinder that holds a rubber blanket. In operation, the plate cylinder is inked and the image transferred to the rubber blanket of the offset cylinder. The substrate is then passed between the rubber blanket and the impression cylinder and the image transferred from the blanket onto the substrate.

All offset presses are rotary and may be either sheet-fed or web-fed. Because of the nature of the process and the use of the offset cylinder, it is impossible to alter the cut-off length of web-fed-stock wrappings. Make-ready is minimal; the wraparound plates can be shifted slightly for proper register. The resilient rubber blanket compensates for the varying thicknesses and textures of paper stocks, largely eliminating a source of trouble in other printing processes. A wide range of papers can be used, and halftones can be printed with solids on both rough and smooth papers, although sometimes it is difficult to maintain uniform color through a run.

DRY OFFSET (COMBINATION LETTERPRESS LITHOGRAPHY)

Letterset or dry offset uses the same kind of relief plates as wraparound letterpress, but instead of printing directly it prints onto an intermediate blanket, as in conventional offset printing. Because the characters on the plate are raised, the dampening operation is not required.

SCREEN PRINTING (STENCIL TECHNIQUE)

The screen process transfers an image through a printing plate rather than *from* one. Developed in 1907, screen-process printing originated in the Fiji Islands, where stencils were first used for cloth decoration. The screen (nylon or fine wire) is porous, allowing ink to pass through onto a substrate. The screen is placed over the substrate, and the ink forced through the screen with a squeegee. Three types of presses are in use today—manual, semiautomatic, and automatic. Each contains the basic elements of a stencil or screen, a squeegee, and some type of back-up plate or fixture to hold the substrate in place.

The manual press consists of a hinged frame holding the printing screen. The object to be printed is held below on the back-up plate, and the ink is applied manually with a squeegee. Both frame and object are held stationary.

Semiautomatic presses can be flat-bed or cylinder types. The simple flat bed uses a mechanical squeegee. In the cylinder press, the substrate is passed under the screen by means of a rotating cylinder. To print cylindrical objects, the cylinder is replaced by the object itself, which rotates beneath the screen.

The automatic machine works on the same principle as the semiautomatic variety, except that the squeegee moves continuously, or, in the case of a stationary squeegee, the screen moves continuously.

The 1980s will see improved color separations, better lenses, and improved inks and presses—all contributing to the sharpest possible detail even on the smallest package.

6   TEST MARKETING

The heightened competition in test markets has become disquieting to many marketing men. They fear it is distorting the sales results obtained in these experiments, thereby reducing the effectiveness of an important research tool widely used by makers of consumer products; far too many products that presumably have done well in test markets are proving to be busts in general distribution, they note, and failures in the test-market phase itself are too frequent. So marketing men are warning manufacturers to read their test-market results with a more critical eye—and to use other methods as well to try to predict a product's reception by the fickle mass public.

Ted Stanton, *The Wall Street Journal,* May 24, 1966

Sampling is an old method for testing a market or a product, but until Perley G. Gerrish developed a program for test-marketing, there was no way to project the success of a new product. In 1882, Gerrish, a distributor of "Squirrel Brand" assorted nuts, came up with a recipe for a new confection, the peanut bar. Carrying a shoebox filled with samples, he called on a number of his regular customers. They refused to stock the peanut bar because they thought that children would never buy it. Unconvinced, Gerrish loaded a wagon with nut bars and traveled from Boston to Providence. At every school along the way, he handed out free samples to the children. A few weeks later, he retraced his route and called on his regular customers again. This time they bought the peanut bars because children had been asking for the new product, pennies and nickels in hand. Gerrish had proved the value of test-marketing.

Today, test-marketing new products is critical to every manufacturer. Now market tests, in-use tests, and other commercial experiments in limited geographic areas are conducted to determine the scope of a manufacturer's marketing program. Design and production factors often must be adjusted as a result of the test findings, and management must decide whether to market the product at all.

Test-marketing must be carefully planned in order to avoid some common problems. Among them:

1. Competitors can take advantage of the time lag between test-marketing and actual marketing to introduce their own version of the new product *without* pretesting.
2. The company may face production or distribution problems caused by limited product manufacture.
3. The company may not allow the test to run long enough to monitor repurchasing patterns.
4. The new product may "cannibalize" an already successful product.

Motivational researchers believe that many test markets are not reliable. They feel that as soon as a product name, design, or package is specifically questioned, the consumer is too conscious of the design and is no longer typical. The consumer reacts to the package rather than to the product. Using the indirect methods of motivation research, consumers' reactions are determined as to packages, ads, brand names, symbols, and designs. A 1980 survey conducted by the Overlock Howe Consulting Group found that 67 percent of their respondents do not conduct any form of consumer research. Nearly three out of four large (over $50 million) corporate marketers tested their packages to minimize introductory risk, while only one in five of their smaller counterparts did this. More than 38 percent of those questioned never evaluated their product packages on the shelf.

Fig. 6.1 *When the aerosol shaving-cream can was test-marketed by Rise in the early 1950s, it rapidly cut into the shaving-soap market. Courtesy, Landor Associates, San Francisco: Museum of Packaging Antiquities.*

The test-marketing of ITT's retort pouch, introduced in 1977, clearly illustrates the pitfalls inherent in a testing program. The pouch was withdrawn within a year because ITT executives claimed that their competitors purchased all their test market samples from retail stores, thus sabotaging the program. No meaningful results were obtained, and the product was taken off the market.

Test marketing will be particularly important in the 1980s. With increased market segmentation, proper market research will be essential to product success. In 1950 the life cycle of a product was 15 to 20 years, in the 1980s it will be less than 3 years. Each year, more than 5,000 new products are introduced on grocery shelves alone—that's some 50,000 during the 1970s and compares to the 8,000 brand items stacked in a typical supermarket.

The term *marketing mix* includes all the factors involved in selling a product. In the case of a new product, the test market tries out *all* the various ingredients in the marketing mix. In the case of an existing product, the test market evaluates a change in only *one* ingredient in the mix.

Some questions that should be answered in a good test program include:

1. What will the market share be?
2. Who will buy the product, how often will it be bought, and for what purpose?
3. Where are the purchases made, and at what price?
4. What changes are being made by competitors?
5. What effect does the new item have on established lines?

TEST-MARKETING NEW PRODUCTS

The aim in test-marketing *new* products is to determine *all* the various ingredients in the marketing mix that contribute to the product's profitability. These include:

1. The product itself
2. The uses of the product and consumer attitudes toward it
3. The *packaging* of the product from a functional and "design" point of view
4. Price structure: trade, retail, and manufacturer
5. Distribution
6. Promotion
7. Media
8. Whether the product is sampled and repeat-purchased
9. Whether the manufacturer's marketing and sales organization is performing effectively.

In other words, testing a new product determines how the product will fare in the marketplace without the expensive risks involved in a full-scale launch. Pretesting something that is to be *sold* is very similar to the more familiar process of pretesting something that is to be *made*. Only about 45 percent of new products succeed in tests. The reasons for the high failure rate include an inferior product, a "me-too" product and poor consumer acceptance.

No product should be test-marketed until it has been perfected and its marketing approach established. All too often, marketing strategies are devised without proper thought and research. If questions are raised about a potential problem, the answer is that it is only a test and all wrongs will be corrected by the time of the national launch. This attitude assumes that the product is the only thing being tested, and that all the other elements that make up its impact on the trade and the public will somehow evolve separately while the product itself receives a fair trial. The later, full-scale launch will be more successful if the alterations required in the product and its marketing are kept to a minimum.

COMPARATIVE TESTING OF ESTABLISHED PRODUCTS

Test-marketing established products is done when a manufacturer wants to change one or more elements in the existing marketing mix. This inevitably involves *comparison* testing. For example, a manufacturer might want to determine whether heavy advertising would profitably increase the sales of his product. He could institute a heavy campaign in one area of the country and compare the results in the rest of the country where advertising continued at the same level. But if he also wanted to learn whether he should use a blue package or a red package, he would have to conduct two tests with the same heavy advertising campaign in each, one with the blue package and one with the red.

Comparative testing gauges the effects in the marketplace of modifying some factor in the product package, such as a design change. When R.J. Reynolds changed the package design of Camels, consumers thought that the cigarettes were also changed. After an unsuccessful test market, Reynolds quickly went back to its original design. In addition, a comparative test provides an assessment of the effectiveness of a change in the product itself. However, the variable under test must be an important one, and it must become apparent during the relatively short period of the test.

PLANNING A TEST MARKET

When the decision is made to test-market a product, a suitable locality must be found that is representative of the larger market. Also, the sales of the product under test should reflect those in the country as a whole. Consumer acceptance of different products varies widely from one part of the country to another. Thus selecting the wrong area for the test could invalidate the results.

The test areas should always be compared with the national market, and regional variations in all facets of consumer usage and attitudes to the product type should be examined, as well as the marketing mixes of competing products.

The reliability and usefulness of the test results will also depend on the amount and accuracy of the demographic, sales, and research data available and the extent to which the test market may be treated as a self-contained marketing unit. It is not surprising that these preconditions are met by only a few available sample markets. They include:

Selected localities. These towns are difficult to define as self-contained, since some spending power in the surrounding countryside goes to the town, some to local stores, and some to other towns. But good results are often cheaply achieved by using towns from different areas in the test. Multimarkets should be comparable in demographics and collectively represent 2 percent to 3 percent of the U.S. population. Fort Wayne, Indiana, has long been regarded as "America in microcosm" because it just about matches the national norms demographically and economically. It has the added advantage of being fairly isolated from "foreign media." The city's two newspapers and three TV stations provide what is considered to be an excellent indication of the effectiveness of an ad campaign. On any given day, Fort Wayne's 25 shopping centers may stock as many as 15 test products.

Television areas. These have many research facilities, but they tend to be expensive. Only a few areas, and not the cheapest, are representative of the country as a whole.

Groups of retail stores. Projecting results up to full-scale market levels are difficult because a group of stores is not a self-contained marketing unit. The advantages are cheapness and the close surveillance of consumer behavior at the point of sale.

Other markets are geographic areas, sales areas, truck delivery areas, and special retail outlets such as supermarkets. But which is the most suitable in a particular case? The planner has to balance a number of factors.

A market can become saturated if the people are aware that they are being tested. If a test is conducted properly, the consumer should not know that he or she is involved. The trend has been to locate mid-size cities that reflect national demographics and use these markets repeatedly but quietly.

The ten most commonly used test markets are:

1. Erie, Pennsylvania
2. Syracuse, New York
3. Binghamton, New York
4. Wichita, Kansas
5. Fresno, California
6. Spokane, Washington
7. Fort Wayne, Indiana
8. Little Rock, Arkansas
9. Peoria, Illinois
10. Portland, Maine

In 1980, Dancer-Fitzgerald Sample added six cities to its annual guide of test markets and dropped eleven others: Added to its list are Buffalo; Evansville, Illinois; Fresno; Little Rock; Minneapolis; and Pittsburgh. Off the list are Austin, Texas; Duluth, Minnesota; Jackson, Mississippi; Lexington, Kentucky; Rockford, Illinois; Greenville–Newbern–Washington, North Carolina; Greenville-Spartanburg, South Carolina; and Asheville, North Carolina.

Choosing the Promotion Method

An important factor in planning the test market is to relate the promotional effort to the national situation. In particular, this means media promotion. If the effect of television advertising is the element in the marketing mix under test, or if television advertising is essential in selling the product, then the area for the test will almost certainly be a television area. There should be at least three commercial TV stations in the test market considered. Fewer stations cannot reliably reflect the views of a cross-section of the population. (Too much cable TV in an area can bias the results.) TV advertising makes the consumer aware of the product, but print media, a brochure, or package copy is required to give the details necessary to convince the consumer and move him to buy.

The Company's Operations

In choosing a test market, the planner has to take into account the special factors peculiar to his company that could influence the test results. Some of these special factors are:

Representation. All sales forces have strong links and weaker links. In small areas, both strong and weak representation can produce incorrect results. An over-zealous salesman may "push a product" too hard, thus causing overly optimistic results. A weak salesman may not attempt to keep the test market stocked with the product, causing false sales figures.

Delivery. Timing and efficiency are particularly important with perishable products. There would be little point in testing in an area where the product will lose its freshness by the time it is delivered.

Performance in key retail and wholesale outlets. This is one of the biggest difficulties in planning test markets because many areas are dominated by major retailers or wholesalers who do not handle the product. Also many major retailers are not enthusiastic about stocking a line in only a few of their shops—those that happen to fall into your test market. It is often advisable to avoid those areas dominated by these large retailers, or at least to take account of their reactions when assessing results.

Background of promotion. An obvious consumer bias can result from heavy promotion or lack of promotion, in the months or years preceding the test in comparison with the national effort.

Historical factors. Many companies find sales difficult in certain areas. This may be due to strong competitive entrenchment or to local consumer preferences.

The Length of the Test

It is a great temptation to accept the results of an initial successful selling to the trade as proof of a successful test market. In fact, a test market must run for sufficient time to measure not only the sell-in to the trade, not only the initial purchase by the customer, but also repeat purchases by the customer for as long as is necessary to establish a pattern, by which time it should also be possible to measure the effect of competitive reaction, if any.

COMPARATIVE MARKETS

If a particular element in the marketing mix of an existing product is under test, the structure of the test plan must isolate this element. The most common example is the testing of advertising to a certain level in one medium, perhaps television or a local newspaper. Here the test area is predetermined by the area of transmission or circulation of the station or paper selected.

Comparative testing becomes complicated when two or more test-market operations are conducted to determine the most preferable course of action: for example, the more profitable of two new prices or the more profitable of two levels of promotion. Two or three test areas are required, and to compare results the areas must be as similar as possible. Ideally, the only differences between the areas should be the element in the marketing mix under test. Two excellent markets with matching demographics are Erie, Pennsylvania, and Phoenix, Arizona. Balanced media markets include Erie and Spokane, Washington. The areas selected should compare well in terms of:

> Population
> Class profile of population
> Urban density
> Trade structure
> Typical total market
> Typical company operation
> Representation and delivery
> Special factors (for example, water hardness or ratio of people
>   living on pensions)

However, such perfection of comparability is difficult to achieve. Some other broad considerations should give a lead to the size of the test market:

If the area or town is small, it will have a low level of company investment, with the possibility of less data being given to competition. In a larger area, it will be easier to watch the true effect of promotion. It will also be more likely to reflect a national operation and the results will approximate those on the national sale.

SOME TEST-MARKETING "DON'TS"

Having planned the method and level of promotion in the test area, the remaining problems of the planning function can be summarized in few common errors:

1. Overattention
2. Insufficient attention
3. Incorrect volume forecasts
4. Establishing unrealistic in-store conditions
5. Incorrect media translations
6. Changing objectives after the results are in
7. Selection of the wrong test markets
8. Failure to take into consideration the atypicality of the test market
9. Assuming the competitive environment will be the same nationally
10. Failure to conduct consumer research

Some of the most important "don'ts" to remember are:

Don't allow the sales force to be overenthusiastic. An area manager far from the head office suddenly faced with the Marketing Manager and Sales Manager running a brand-new line on test may often over-promote the product in a manner completely inconsistent with that planned if the product achieves national distribution.

Don't overpromote. There may be only one or two key customers in a test area, and the cost of supplying these customers with the test product may seem quite small, but again it could not be repeated on a national launch.

Don't use special products or packaging. A test market is just as much a test of production and packaging capabilities as it is of sales and marketing. The product should come off the machines exactly as it would in national distribution, and it should be packaged in the same materials.

Don't let the test run into seasonal unbalance. Although the test may have begun at a normal time of the year, it could easily be thrown off if it is allowed to run into a festive season or an off-period for the product.

Don't forget that your competitor is also testing. He may be out to undercut you, and he is not bound by any of the rules that must be observed for a true test market. Competitive action must be watched closely, and if it's excessively heavy or excessively light (for a competitor may not bother about your small test market but might certainly react to a national effort), then you must make allowance accordingly when you evaluate the results of your test effort.

Test-marketing is a viable tool for the sales executive, if he selects a valid test market, sticks to his original objectives, keeps the product in test long enough to obtain repeat purchase information, and most important of all, looks at the results with more candor than optimism.

Fig. 6.2   *Food in metal tubes has never been successfully test marketed in the United States. Courtesy, Metal Tube Packaging Council of North America, New York City; photograph, Camera Arts Studio.*

7 PACKAGING ECONOMICS

The bitterness of poor quality remains long after the sweetness of low price has been forgotten.

Anonymous

Sometimes products are packaged in materials that cost more than the product itself. The superb Huntley & Palmers biscuit tins of the early 20th century cost many times more than the product, and the amount of work involved in the production of these tins was enormous. In many cases, the containers were specially commissioned, but they could also be chosen from the printers' annual catalogue, since many tin box printers offered standard designs that could be chosen by the box user.

Usually, the product category or commodity determines the package material cost. Expensive appliances, such as refrigerators, dishwashers and ovens, are often packed at an outlay of less than 5 percent of their total production cost. But other important factors are included in the overall picture, such as developmental costs, the selling environment, production factors, distribution costs, storage/handling charges, and losses that result from damages, returns, and products that remain in the distribution pipeline once a package change occurs.

Fig. 7.1 *Huntley & Palmers biscuit tins from the company's 1909 catalogue. The small powder compacts would have contained iced gems. Courtesy, Associated Biscuits Ltd., Kings Road, Reading, England.*

Many people consider the package to be a product. The consumer wants value for his money, and often he will pay gladly for the combination of quality and convenience. Salt in a ready-to-use paperboard shaker costs much more than salt bulk-packed in a carton, yet the consumer likes the convenience of the table-ready package. Individual-portion jams and jellies are now being sold in many supermarkets. The convenience of use often precludes the product's high cost, as opposed to a glass jar of jam.

DEVELOPMENT COSTS

Packages are developed and formalized in a series of costly operations, and very few packages escape this time-consuming process. The steps involved in package development include design, prototype construction, test marketing, specification development, quality-control implementation, and initial start-up. The highly touted retort pouch had an enormously expensive development cost of more than $5 million, while a candy that is dump-packed in a polyethylene bag has a negligible development cost.

In the packaging industry the material supplier (or converter) usually absorbs the initial material development cost and then spreads it over several years of production. A supplier also may include development costs in an overhead figure and spread it over all the materials sold. For example, the last several years have seen a noticeable decrease in the price of polyester films because of significant improvements in manufacturing and the initial development costs that are already being absorbed by the producer firms.

Designing the package is an important part of the manufacturer's development costs. Manufacturers can hire outside designers, tackle the problem in-house, or use the services of a supplier and designer.

Many large product manufacturers employ one or more staff designers. Normally, their responsibility is to act as a liaison with outside design firms and possibly to work on mutual projects. In-house design staffs are favored by many large corporations because they can retain control over the final designs and eventual cost.

Outside designers are often brought in when a broader range of creative alternatives is needed. Objectivity allows them to challenge management for the outstanding package design. Often an outside designer's fee is negligible in comparison to the total advertising appropriation. A workable figure for most packaging design projects is about 10 percent of the total advertising budget, which in itself could amount to about 5 percent of gross turnover. This figure is one-half of 1 percent of the gross turnover.

Edmond A. Leonard in *Introduction to the Economics of Packaging* lists the factors in the costs of developing a package:

> Identification of package criteria
> Concept search
> Design
> Models
> Sample tooling and samples

Sample evaluation (technical and customer research)
Costing and specifications
Tooling and materials for test market
Test-market pack and evaluation
Specification refinement and purchasing
Tooling for production
Quality-control implementation
Start-up

Although not every package is subject to all these costs, the product manufacturer must be aware of these factors and plan his product prices accordingly.

Using a Design Firm

If a manufacturer feels a risk is involved in hiring an outside designer, he can write a break clause into the contract. Assuming that an entire package design project will cost $10,000, the break clause in the contract might state that if the client is dissatisfied,

Fig. 7.2   *The Merolite one-trip container was a 1975 World Star winner for an unusual soft-drink container. Courtesy, Plastics Division, Imperial Chemical Industries, Ltd., Hertfordshire, England.*

at the end of the first presentation stage only half the total fee would be due. This clause also protects the designer from performing work not originally specified by the manufacturer.

Outside design firms might charge a client on an hourly basis or on an accept/reject basis. Or the manufacturer can retain a designer for a fixed amount per year. In this fee arrangement, the manufacturer pays a fixed amount for exclusive service in the specific product area as well as for each design job at conventional rates. In the United States, many professional package designers are members of the Package Designers Council (P.O. Box 3753, Grand Central Station, New York 10017), an organization founded in 1952 by industrial design consultants who specialize in package design.

One disadvantage in using an outside design firm is that it may specialize in only one area, such as ethnic, nostalgic or ultra modern designs. But this can also be an advantage, since the firm can handle special projects.

Other Design Sources

Three additional design sources are available to the product manufacturer at varying fee structures: advertising agencies, free-lance designers, and package suppliers.

An advertising agency often provides package design services to a client, even though it is not its main business. Design work by an agency may be done either by a staff designer or an outside design consultant brought in by the agency. The main purpose of an advertising agency is to create advertisements, and practically all departmental functions are directed toward achieving this single goal. Often an advertising agency can better integrate a certain package in its campaign by designing it in-house. Advertising agencies are usually weak in the technical aspects of packaging and sometimes fail to understand the basic differences between materials, shape, and shelf life. But often excellent designs can be obtained by combining the creative aspects of an agency with a firm knowledge of materials and package requirements. An advertising agency's fee for design work is fairly high.

A product manufacturer can also hire a free-lance designer or use a package supplier's firm. A free-lance designer often works for a wide variety of industries. Successful industrial designers such as Raymond Loewy and James Pilditch have designed many other products in addition to working on successful package design.

Most major packaging-material suppliers employ a design department as a customer service, usually at no charge to the customer. The cost of the design is absorbed in the overall material cost and supplier's overhead. Using a supplier's designer is probably the most economical initial route for a product manufacturer—if he is aware of the limitations. Obviously a supplier's creativity often is limited to materials that are sold by the supplier, and their design work can be narrow in scope and limited to the specific problem of the material order.

A survey on package design attitudes conducted by the Overlook Howe Consulting Group in St. Louis rated package design sources and found the following design preferences:

Source of Design	Preference (Percentage of Respondents)
Corporate staff	41
Package design firm	38
Advertising agency	26
Freelance designer	23
Package supplier	5

PACKAGE MATERIALS COSTS

Suppliers will quote costs on either a pound, area, or unit-pack price. The price for the material is known as the internal cost-per-unit pack. The actual cost to the user of the packaging material is obtained by calculating items such as volume, special packing, freight, and storage and handling.

Of course, the larger the order the greater the savings on material cost. Packaging material suppliers often offer price breaks at certain levels. This takes into account start-up costs for the material in the supplier's plant. The quantity of the order also may dictate the specific manufacturing operation. Extrusion favors long runs since the start-up scrap is substantial. Long runs are also more economical when printed by gravure instead of

Fig. 7.3 *The Drain Power aerosol container was a 1975 World Star winner for the United States. The package depicted is the original packaging and is no longer current; it was discontinued in January, 1978. Courtesy, Airwick Industries, Inc., Carlstadt, New Jersey.*

flexography. The cost of an etched gravure roll favors longer runs. Since flexography uses rubber plates, its initial costs are less, and smaller runs are more desirable. There is a break-even point above which gravure printing gives a lower unit price than flexography. This point depends on the material specifications and on the exact type of equipment used.

Most packaging material is delivered to the user on large pallets that can be stretch-wrapped or wire-banded. Bulk palletization is economical for both the supplier and user. Some users might specify that the material be individually shrouded in polyethylene or inserted into separate corrugated cartons with special core labels. When special packing or identification is requested, costs increase.

Shipping costs are expensive. Air shipment or ordering minimum quantities often bear upcharges to the materials user. The unit price is affected by the distance between the user's plant and the supplier. If returnable packing such as pallets, is used, the supplier may collect these at a regular schedule. Aluminum cores for flexible packaging rolls are also returnable. When these are shipped to the supplier's plant, credit is given to the user for their safe return.

Rarely does an order arrive at a user's plant without any intermediate storage or handling. A supplier may store the user's material against a release for delivery in a rented warehouse or a locally owned warehouse. This service costs a premium to the materials user and is usually reflected in the initial material-cost quote.

PRODUCTION COSTS

Production costs include all the operations required to produce the final saleable package. They include the fixed costs, such as labor, machinery depreciation, and overhead, and variable costs, including the cost of the raw materials and packaging.

One technique that can lower production costs is to accelerate the modernization of the production processes. The packages should present a quality image and be able to be made on high-speed packaging machinery. Fixed costs can be lowered by more efficient machine utilization or savings in labor. Variable costs are reduced by savings in raw materials or packaging.

Machine efficiency is a critical factor in determining the financial success of a finished package. Consider two types of cartons—glued and unglued. Assume that Machine A produces unglued cartons, while Machine B produces glued ones, and note the cost differential in this example:

Cost	$30,000	$20,000
Speed (pkgs/min)	100	150
Floor space (square feet)	400	300
Labor	1 operator	1 operator
Material cost/1,000	$15.00	$19.00

Additional factors to consider include the cost per hour for the different jobs per machine. After the final calculations have been made, the unit package cost for Machine A is less than Machine B when one million or more cartons are run, despite the lower cost of Machine B and its 50-percent-greater production speed. The predominant factor is the cost of the packaging material. Here

Fig. 7.4 *The Model CMH Cartoner with vertical polycarbonate barrier and emergency lifeline over bucket conveyor. Courtesy, R. A. Jones & Co., Inc., Cincinnati.*

the economics preclude the production speed and machine efficiency. Changes in packaging-material specifications also directly affect overall packaging-machine efficiency. Samples should be run on a trial basis to ensure that the production order will be economical.

Labor costs must be calculated when the packaging operation is not automatic or even semiautomatic. Direct labor costs are the wages paid to the machine operators and other personnel handling the packaging materials from the beginning of production to the time the finished package is shipped from the plant.

Effective supervision also plays a part in high production rates. Often incentive bonuses based on production are given to increase production levels; while this is good, quality could suffer in the process.

Packaging losses, or shrinkage, refer to quantity rather than size; they are the difference between the units of packaging material received and units of packed goods shipped. It is a vital factor in overall production costs. Some loss is unavoidable, of course. Large losses can occur through failing quality-control standards and during machine start-up either at the beginning of a shift, after a break, or after a size changeover. Down-time as a result of machine or product problems also leads to losses.

Accounting allows for a standard shrinkage factor, and this varies from material to material. For corrugated cartons, it is usually less than one percent, while for flexible packaging materials, the value can range as high as 5 percent, even for an efficient operation. In a flexible-packaging operation, potential losses may occur during the initial quality-control inspection, machine start-up, filler adjustment, line samples, off-quality production units, splice down-time, and material remaining at the end of the core. Unusual problems can play havoc with economics and greatly multiply the shrinkage factors.

All these losses are small if considered individually, but they can involve significant amounts of money over a period of time.

WAREHOUSING COSTS

Just as the materials supplier stores his material, the package manufacturer normally ships the packed product to a warehouse for storage prior to final distribution. Certain products have specialized warehousing conditions that increase the cost of the final unit. Frozen foods must be kept frozen at all stages of storage and distribution. For certain frozen fruits and vegetables with short harvesting periods, enough packages must be kept frozen to fill demand on a regular basis. Confections containing chocolate require controlled-temperature storage in the summer. Refrigerated and freezing warehouses can be rented in all major cities.

Additional factors that come into play in calculating warehousing costs include:

1. Modifications in filling, storing, or shipping
2. Physical movement from filling line to initial storage
3. Inventory coding with the use of a Universal Product Code on the package.

Package shape directly influences storage costs, since it affects overall space utilization. This becomes important in the case of

Fig. 7.5   *The Wita BoxBrot retail package overcomes many of the disadvantages of conventional bread packaging. The polypropylene box is biodegradable; the two-package unit is practical and hygienic, since half is open while the other half remains sealed. The unit pack is connected with aluminum foil, which is hinged and can be reclosed easily due to a specially constructed closure in the box lip. BoxBrot is pasteurized and, in the closed pack, has a shelf life of many months without additives or preservatives. Courtesy, Bolletje A. A. Ter Beek B.V., Almelo, West Germany.*

Fig. 7.6   *See-through plastic resins permit the production of appealing lipstick packages. This pack uses acrylic plastics. Copyright © 1977 S.E.C. Ltd.; Courtesy, Sinish Communications, Westport, Connecticut.*

frozen foods, where low-temperature storage is expensive. Package strength is another critical factor that affects the costs of storage. In single-story warehouses, products can be stored either in racks or dead-stacked in cartons up to the roof. This requires that the outer pack and its contents stand up to large stacking heights. If the more expensive method of rack storage is used, the outer carton strength is not important.

TRANSPORTATION COSTS

The costs of shippers and shipping are determined according to various freight-cost schedules published by the industry. No advantage is gained by using inferior materials if the product arrives at the retailer damaged. Rule 41 of the Uniform Freight Classification defines the minimum acceptable standards for corrugated boxes to be used for rail shipments. Presently, nearly all corrugated shippers must meet Rule 41 or pay scheduled upcharges for freight tariffs. In addition, a common carrier has the right to refuse to handle products packed in a container that does not fulfill the requirements of Rule 41.

Freight rates are also fixed by commodity according to its value, the average density in pounds per cubic foot as shipped, and the distance involved. Lightweight products have higher rates than dense ones.

TERMINAL INVENTORIES

No package lasts forever. When the time comes for a package redesign, for whatever reason, no doubt a considerable supply of old packages will be in stock. Should these be sold to a close-out operation or left on the shelf? Can the manufacturer afford to leave old and obsolete packages in the distribution network? These are hard questions, but every packer of large-volume products must answer them.

The problem of terminal inventories is further complicated by ever-changing governmental regulations. Old, obsolete labels may have to be discarded, as may an already packaged product.

New label requirements, such as more complete nutrient information, often demands the printing of new labels. The cost to the product manufacturer may become staggering

PROMOTIONAL COSTS

An indirect factor affecting packaging costs is the cost of advertising. This may exceed the packaging cost for many goods, and at times packaging costs may be reduced to provide more money for sales promotion. This could lead to underpackaging, however, and the best distribution of expenditure has to be decided to ensure both good packaging and adequate advertising.

8 ROLE OF THE PACKAGE IN PRODUCT SUCCESS

If you want me to do a big thing like a tractor, there are so many obvious things you could do to make it better looking that I would charge very little. But if you want me to redesign a sewing needle, I'd charge $10,000. After all, how can you improve a needle?

Raymond Loewy, Industrial Designer

The package contributes to the marketing success of a product in two areas: in its ability to sell the product and its ability to contain the product for a satisfactory shelf-life. Ignoring the package often means both profit loss and product failure in the marketplace. Packaging is not an end-of-the line wrapping-up process: It must be fully coordinated with product design and be recognized as a vital factor that influences sales, costs, and profits. Many test methods are used to determine the consumer's attitude about packages and their design. The most common ones are listed in Table 8.1.

The package must be developed to suit the product. In Great Britain many people associate fish and chips with a newspaper wrapper, so some manufacturers developed a waxed-paper chip cone printed in traditional newspaper style. This allowed the consumer to associate his new hygienic paper cone with the "wholesomeness" of the newspaper cone of past years. It became a quite successful package because of the increasing interest in cleanliness by the British consumer. Most packages are tailor-made for the product and intended only for that specific item. This is either because of technical requirements or design criteria.

"Silly Putty," a children's toy made of a silicone derivative, was originally packaged in plastic eggs. The eggs were economical and easy to ship in inexpensive egg cartons, acquired from the Connecticut Cooperative Poultry Association. Later, when the makers of Silly Putty reexamined their packaging and tried another container, top toy buyers protested loudly, and the original egg package was reintroduced. The brightly colored eggs seemed to imply that the toy inside was fun. This egg-shaped package probably has been responsible for the success of "L'eggs" panty-hose. It introduced the consumer to the idea that a plastic egg can be used to contain a saleable product.

Package shape has proven to be an asset in the sale of products such as Janitor-in-a-Drum, Fantastik spray cleaner, and Coca-Cola (often considered to have been designed after the shape of a female). A unique shape can be an attention grabber, reinforce the product position, or even act as a trademark.

In the technical area, packaging materials specifically developed for one product often cannot be used for a different product. Even the same product, such as ketchup or mustard made by different manufacturers, sometimes cannot be packaged in the same material, due to differing product formulations that may affect the shelf-life of the product.

PACKAGE GRAPHICS

The importance of package design as a marketing aide can be illustrated by several case histories from the beverage industry. The success of Kirin beer and "Pockit" fruit drinks was enhanced because the companies developed a saleable package. The makers of Chelsea soft drinks, however, failed to gauge the market accurately. And a British case history, of Rowntree's "After Eight" chocolate mints, illustrates how an advertising agency turned a discernible change in public taste into a marketing success.

TABLE 8.1 METHODS USED IN PACKAGE-DESIGN TESTING

Test	Object	Instrument Used
Visibility tests	Visibility at distance	Distance meter (Polarascope)
	Visibility at angle	Angle meter
	Visibility as function of lighting	Illumination meter; "Diaphragm glasses"
	Visibility under poor optical conditions	Use of filter
	Visibility for short time span	Tachistoscope
	Sharpness of package	Threshold of perception meter
Eye-movement test	Traces eye flow and shows where attention is held. Shows how design guides the eye.	Eye-movement test with an eye-flow recording instrument Eye-camera
Readability test	Determines the readability of the brand name and product from the shelf or freezer.	Tachistoscope Visuometer
Controlled association test	Consumer attitudes toward marketing tools (motivation research technique)	Tests with control groups
Focus group interviews	Consumer attitudes toward marketing tools (motivation research technique)	Depth interviews with sound film or videotape

Test	Object	Instrument Used
Semantic differential or polarity test	Tests certain package or product characteristics.	Questionnaire
Texture scale	Tests texture based on visual aspects of package design.	Questionnaire
Physiograph with pushbutton	Rates phrases using conscious and emotional responses.	Physiograph
"Reisco" (emotional) response index system	Copy pretesting system. Uses a data bank with emotional responses of 7,000 consumers to stimuli.	Computer
Apparent size of package	How package looks and appears on shelf.	Tray/package/tunnel set-up; Psychomagnometer
Localization of package	How package is noticed among others on shelf. Measures attraction-power.	Mass display impact test (use of slides)
Unity of design	Analysis of package as a graphic unity.	Torsion stereoscope

Kirin Beer

Although Kirin beer is the largest selling beer in Japan, it had until recently marketed its beer in the United States solely through restaurants. The company's decision to go after American grocery sales coincided with strict labeling laws and Kirin's use of a new, more elongated, nonreturnable bottle. The new shape provided a stronger, lighter bottle, which became a substantial benefit for an export beer. It was necessary then to redesign the neck label in order to list the ingredients and find a shape more compatible with the taller, narrower bottle.

Kirin also markets a premium-brand beer in Japan called Mein Brau. Extensive studies showed that waiters and waitresses disliked the old heavy-foil neck labels; because of their stiffness, it was difficult to remove the cap quickly. The manufacturer decided to redesign the neck label. Later, the main label was also redesigned, reducing the number of typefaces from six to one, refining some details, and changing the emphasis from "brewed in Japan" to "Kirin Beer." The drawing of the kirin, a mythological creature resembling a combination of horse and dragon, was also clarified.

Since it was important to preserve the Kirin symbol but not detract from the main label, the neck label, after numerous trials with colors or with colored bands on gold, was a deep band, silver-foil-printed matte gold with a narrow, glossy gold border framing the slightly pointed bottom. The small Kirin symbol was printed in red, and the brand motto, "The Beer of Legend" under it was printed in green in specially designed script. The rounded main label was lithographed on silver foil in white, red, green, ocher, and glossy gold.

Because of this extensive redesign for a newly emerging market, Kirin noted a 30-percent sales increase on the East Coast during the first year and a 50-percent jump on the West Coast and in Hawaii.

Shasta's "Pockit" Fruit Drinks

For "Pockit" fruit drinks, a newly developed package concept was combined with nostalgic graphics. The package and the product itself were new. Introduced in 1979 by Shasta Beverages, "Pock-it" was packaged in the German "Doypack." Composed of an aluminum foil laminate, the Doypack is a stand-up pouch that is widely used in Europe. Shasta Beverages purchased the American distribution rights for the package.

For graphics, an old-fashioned, full-color "orange-crate" look was used, combining whole fruit, shiny leaves, and a juicy looking half of a cut lemon, orange, or apple, which emphasized the drink's wholesome quality.

The novel shape of the pouch gave the designers a larger front to work with than a traditional beverage can. Although it sits on a firm, oval-shaped base, like a can, the pouch tapers to a flat, horizontal closing that exposes more of the package to a frontal view. A drawing of the package itself with a straw protruding appears on the front panel, creating a mirror effect and showing where the special pointed straw should be inserted above the liquid line. Use instructions and ingredients on the back of the pouch are scatter-printed in bright blue. Pouches are sold in six- or twelve-packs with the straws.

The graphics are printed in Germany and the pouch is made in the United States. The graphics are printed gravure and run backwards and upside down on a clear film. The film is laminated to the aluminum foil, and the laminated sheets are then trimmed into long strips. At the bottler, the strips are cut, heat-sealed into bottle shapes, filled, and sealed.

The package appeals to the consumer because there is no spill and no breakage and it is lightweight and easy to store. The advertising agency for Shasta reported that "Pockit" achieved 85-percent purchase in the Yankelovich market test, the largest percentage ever recorded for a new beverage.

Anheuser-Busch's "Chelsea" Soft Drinks

Introduced in 1977 by Anheuser-Busch, "Chelsea," a lemon-lime-ginger-apple beverage, was made with all natural flavors and with about one-third less sugar than conventional soft drinks.

Packaged in a "Michelob"-type bottle with similar foil neck and body labels, the product was created to fill an existing product void—an adult alternative to conventional soft drinks or wine. Most soft drinks contain a small amount of alcohol, but Chelsea contained more because natural flavors were used.

The product turned out to be a multimillion-dollar failure. Critics contended that Chelsea was Anheuser-Busch's attempt to lure children and young adults into the beer habit. Its beerlike bottle did not suggest a soft drink to the consumer. Anheuser-Busch attempted to line-extend into a semi-soft-drink product, but its identification as the manufacturer of a successful beer doomed its efforts to failure, despite superb package graphics and the use of aluminum-foil labels.

Rowntree's "After Eight" Mints

When the J. Walter Thompson Agency was hired to promote a new range of chocolates by Rowntree, they chose "After Eight" for their promotional thrust.

The new mints in their individual paper sleeves stood at their best as a luxury product that would adorn a polished dinner table. "After Eight" deliberately invoked this feeling of leisure elegance, which was matched by the design of the box and the motif—an ormolu clock with the hands appropriately set. In a national promotion, this appeal to status was given free rein: Black ties, vichyssoise, candlesticks, and zabaglione were used to evoke the feeling of luxury. The product has been a huge success.

PACKAGE SHELF-LIFE DEVELOPMENT

There are numerous examples of how the package has contributed to the marketability of a product: Grolsch Lager Beer from Holland with its superb bottle and old-fashioned wire clasp and ceramic stopper; Listerine with its olive brown outer paper wrapper; Crosse and Blackwell's Major Grey's chutney (domestically made but carrying an imported stance); and the beautiful decanters offered by Schenley's J. W. Dant whiskey.

Proper material selection coupled with machinability often goes a long way toward achieving a satisfactory product shelf-life. One of the most interesting concepts to appear recently is "hybrid packaging," expounded by celebrated designer, Walter Stern. In Stern's view, this is an emerging technology, involving the grafting of one packaging form upon another. The result—a totally new style emerges.

Examples of hybrid packaging include pouched milk and the mold/tray combination.

Milk can be packaged in plastic film or in laminated pouches. The packaging of milk in pouch form was pioneered in France, and subsequently it has been introduced to several other European countries. In Canada, more than half of the population buys its milk in pouches. The pouches are opened by cutting off one corner, the milk is then stored in a molded polypropylene plastic container. Pouches offer economy, compact storage and ease of disposal. Disadvantages include the need for support and an unconventional appearance.

In a mold/tray combination, a large thermoformed plastic tray serves as a container into which the filling mixture is poured. The filling hardens, a printed paperboard lid is heatsealed to the large multicavity tray and is subsequently cut into single-bar packages. In this case, the package serves many functions—in production, distribution, and selling by using thermoformed plastic to form and contain the product and coated paperboard to protect it and display it.

CONSUMER-BEHAVIOR TESTING

The range of packaging materials available today is so wide that very often the manufacturer has considerable difficulty in deciding which packaging system to select for the product. Tests show that the same product packaged differently is thought of by the consumer differently. Package materials also influence the consumer's attitude about a product, and careful consideration must be given to this package when a product is in the developmental stages. Many products have succeeded or failed because of the packaging material. Ice cream in a round polyethylene container has a different appeal than ice cream in a rectangular paperboard carton. It's easier to open and close and offers luxury appeal to the consumer. Another example is that an aluminum foil label denotes luxury, even though a paper label is all that is necessary from a manufacturing level. All products, like living creatures, go through a life cycle. It is essential that the corporate marketing department be sensitive to the position of each of their products in relation to its life cycle. They alone must decide whether the product is showing healthy progress or whether it should be discontinued or revamped.

Too many marketing programs see the consumer as a simple-minded person. In *Business Without Gambling,* Louis Cheskin found many commonly held notions about packaging to be false.

1. *An effective package is primarily a symbol, not a work of art.* An effective package attracts attention, communicates the product, and motivates the consumer to buy it.

Fig. 8.1 *The use of an Indian on this Dutch Masters nostalgia tin conjures up a feeling of the late 19th-century cigar store. It is Dutch Masters Panatela antique cannister, containing 25 cigars and designed for holiday giving. Courtesy, Dutch Masters Cigars and David, Oksner and Mitchnek, Inc.*

TABLE 8.2   PRODUCT LIFE CYCLE (PLC)

Stage	Characteristics
1. Introduction	Product is developed and introduced by product managers, sales, finance, and engineering personnel.
2. Growth	Rapid consumer acceptance and sales increase.
3. Maturity	Predictable sales volume and leveling of sales occurs.
4. Decline	Consumers tire of product, decline sets in, and product is terminated.

2. *Consumers do not resent packages because they think they have to pay for them.* In an affluent society, even those with low incomes want psychological satisfactions, straw-covered wine bottles, more expensive than plain glass bottles, give the feeling of real care and naturalness.

3. *The widely advertised product does not always outsell the unadvertised product.* An effective package will bring more sales than a weak package that is heavily advertised, provided the package has equal display with the heavily advertised brand.

4. *A package in many colors is not necessarily more effective than a package in one or two colors.* Full-color printing presents a natural effect, but the package should be identified by one major color.

5. *A "modern" package will not necessarily increase sales.* Some products will sell better if they are associated with old-fashioned packages. Loose tea sells better packed in a metal box than in a paperboard carton. It is perhaps unfortunate that there are so many psychological, sociological and economic theories concerning human behavior. Marketing is confronted with a great heterogenity of consumer choices and of the factors behind them. It is really in a position of conflict between all these many theories.

MOTIVATIONAL RESEARCH

"Motivation researchers," says Louis Cheskin (Business Without Gambling, 1966), "are interested in what people do and why they do it, not in what they say." Borrowing from the behaviorial sciences, such as clinical psychology, motivational research is designed to reach the subconscious mind of the consumer to determine how he or she makes product choices. By using unconscious-level testing, such as association, indirect preference, and retention tests, results are obtained. Popularized by Louis Cheskin (Menlo Park, California), Dr. Ernest Dichter (Croton-on-Hudson, New York), and Burleigh Gardner (Chicago), motivational research seeks to promote a sharper focus in consumer attitudes toward the products and services they use. Dr. Ernest Dichter advocates depth interviewing, and his book *The Strategy of Desire* has become a primer for later practitioners. Louis Cheskin places less emphasis on depth interviewing and instead has developed a series of controlled tests that are conducted on a subconscious level.

Motivational researchers believe that the average consumer sees the package only as a representation of the product at the point of sale. She is not conscious of the fact that she is influenced by the package, that she transfers the sensation, the imagery, or the color to the product. Cheskin uses the term *sensation transference* to denote the transfer of a consumer's feelings from the package to the product. To reposition Marlboro cigarettes from a predominately female-used product (first cigarette with a cork-tip filter) to a male-used product, Cheskin Associates introduced a crest into the package design. More than 71 percent of those tested preferred the package with the crest. The crest was associated in the consumers' minds with quality and prestige. (Note also the Mercury Marquis and the Marquis Brougham with a crest.) Another product that was tested through the techniques of motivational research was Betty Crocker cake mixes. In the early 1950s, General Mills felt that the Betty Crocker brand, with an oval as the logo, lacked a firm quality image. By changing the symbol of the oval on the package to a "spoon" symbol, sales quadrupled in one year. Consumer tests conducted on an unconscious level revealed that the red spoon produces very favorable reactions. It appears to represent fullness and wholesomeness.

Corporate identities have also been tested by motivational research. Before it was acquired by Arco, Sinclair Oil Company used a dinosaur as its corporate symbol. But tests showed that the dinosaur suggested slowness, sluggishness, and lethargy to the consumer. Compare this with the success of the "Tiger in the Tank" symbol of Esso. The trademark of Standard Oil of Indiana passed motivational testing with high ratings; that firm has used a torch and oval in red, white and blue quite successfully over the years. Tests conducted on cold-cream jars showed that women preferred the jar with a rounded triangle on its label rather than the one with a sharp pointed triangle. Sharp points usually are unappealing on package designs, except possibly for hardware products. The "Mother Nature" logo on the Chiffon margarine container was unpopular with the consumers tested, as was a sharp rectangular design on the Saran Wrap carton. When Jeno Palucci of Chun King fame introduced "Wilderness"-brand fruit filling, he added his name "Jeno's" to the label. Motivational research testing showed that consumers associated his name on the label with a greasy, spicy, pizza-like product. This characteristic is called *element isolation*, and is an important one in the mind of the consumer. Consumers tested felt that the black and green colors used on the Irish Spring soap carton were not effective for soap (black to clean up?), as was the repetition of the name "Libby's" on its line of canned products.

Often there is a definite conflict as to which color should be used for a certain product. Should the colors "jump off the shelf" or identify themselves with the product contained in the package? Motivational research testing can provide these answers to the product manufacturer. In a study of liquid antacids packaged in a glass bottle contained in a printed paperboard carton, two cartons were evaluated—one printed overall red, the other printed overall blue. Although the soothing blue carton was obviously more identifiable with the product, the red carton tended to strike the consumer's eye. A study of bubble-gum wrappers proved that the package presently on the market was acceptable. The test evaluated printed red, violet, and white gum wrappers as to display effectiveness and flavor retention. The consumers tested felt that the red wrapper was not only visually effective but appeared to suggest flavor retention over the longest time period.

TABLE 8.3 SOME POPULAR FOODS AND BEVERAGES: HOW THEY BEGAN

Product	Package	Originator and Date	Background
Cracker Jack	Foil-wrapped board box, printed offset litho	F.W. Rueckheim, 1872	Introduced at 1893 World's Fair. Named in 1896. Jack, the little sailor, and his dog Bingo appeared on the package in 1919.
Popsicle	Paper, printed two colors	Frank Epperson, 1905	Invented by Epperson when he left a mixture of powdered soda mix and water on a back porch. Named "Epsicle," later became "Popsicle."
Barnum's Animal Crackers	Recycled paperboard, printed four-color offset litho	Adapted from traditional English crackers, 1902	Named for circus impressario P.T. Barnum. First appeared as Christmas pack.
Oscar Meyer Weiners	Plastic laminate package, heat-sealed paperband printed by color offset litho	Oscar F. Meyer, 1883	The term "hot dog" was coined in 1901 at the N.Y. Polo Grounds by cartoonist Ted Dorgan.
Tootsie Roll	Paper, printed in two colors, with inside board	Leo Hirschfield, 1896	Austrian immigrant Hirschfield made candy from a recipe he brought from Europe and named it after his little girl, "Tootsie."
Life Savers	Foil and paper, gravure printed	Clarence Crane, 1913	First known as Crane's Life Savers, later sold to E. Noble and a partner. He first used foil wrapper and placed product next to cash register in stores.

Product	Package	Originator and Date	Background
Campbell's Tomato Soup	Steel can with ¼ lb. tin coating; four-color paper label	John T. Dorrance, late 19th century	Originated concept of condensed soup in 1897. Red/white label suggested by colors of Cornell College.
Hires Root Beer	Glass with metal closure	Charles E. Hires, Sr. 1876	Pharmacist C.E. Hires created drink from root, bark and berries. Product called root beer to appeal to beer drinkers.
Hershey's Milk Chocolate	Unprinted inner wrap, label printed one-color gravure on coated 55-lb. paper	Milton Snavely Hershey, 1895	In 1812, Hershey, a caramel maker, first thought of mass producing an inexpensive chocolate bar. He attempted to copy Swiss techniques.
Chef Boy-ar-dee Spaghetti and Meat Balls	Steel can with paper label printed four-color offset litho	Hector Boiardi, 1929	Chef Boiardi created product at his Cleveland, Ohio, restaurant. Operation grew from a little room to large firm.
Coca-Cola	Glass with metal closure	Dr. John Syth Pemberton, 1886	In 1889, Chandler bought product from Pemberton. Later sold to Thomas and Whitehead, then to Woodruff.
French's Pure Prepared Mustard	LDPE, printed gravure three colors	George J. French, 1904	French's product first sold in 1904, the same year hot dogs were introduced at St. Louis World's Fair.

In an article in *Advertising Age* (October 20, 1980), Burleigh
Gardner mentions the reasons why consumer research has not
become a true science:

> First is the fact that almost all consumer research is proprietary and
> is aimed at short-run problems. Second is the complexity of the
> beast. We are in a field more comparable to the medical sciences,
> since the subjects of research are living beings, and as living beings
> they are constantly adapting to what goes on around them, and are
> able to control their actions as consumers. The physicist deals with
> entities that obey stable laws; whether astronomical entities or
> subatomic particles, they behave predictably, they do not have
> minds of their own with control of their own behavior.
>
> Given these two factors, it is not surprising we have had trouble
> developing adequate concepts needed to create a real science in
> research.

POSITIONING THE PRODUCT

Positioning was first used in the advertising industry in the late
1960s. Developed by Jack Trout of Trout and Ries Advertising, it
rapidly became one of the most talked-about marketing and advertising strategies of the 1970s. The concept deals with not what
one does with a product, but what one does to the consumer's
mind. The four steps to successful positioning are:

1. Determine present position
2. Determine position needed
3. Determine competitive positioning
4. Determine financial position to perform job

One of the first to employ this new concept was 7-Up soft
drinks. The top sellers in soft drinks were both colas, Pepsi and
Coca-Cola. 7-Up positioned itself against them as the "Un-Cola."
Sales rapidly increased, and soon 7-Up became a strong third in
the soft-drink market, as well as the leading "Un-Cola" soft
drink. Instead of being number one in a minor market ($54
million in sales), it is now number three ($167 million in sales)
among all soft drinks. Other early examples included Schaefer
("The one beer to have when you're having more than one")
positioning it as the beer for heavy beer drinkers. Avis Rent-a-

Car used an "against" position to battle Hertz and became a widely quoted success story as number 2.

Basic to successful positioning is finding out exactly where the product fits into the consumer's thoughts and behavior. In 1969 RCA tried to enter the computer field but failed ($250 million loss in 1971), even though its products were technically superior to IBM, because the company did not take advantage of its firm market as a manufacturer of communications equipment. The consumer felt that IBM was best in computers.

There are also many successful examples involving repositioning the competition. Royal Doulton, which is made in Stoke-on-Trent, England, in competition with Lenox, which is made in Pomona, New Jersey, told the consumer that it was better because it came from England. Vodka is perceived as a Russian product, and the Smirnoff ads and package designs give this impression, even though it is made in the United States. But Stolchinaya, made in the U.S.S.R., capitalized on this fact, advertising "Most American vodkas seem Russian. Stolchinaya is different. It is Russian." In both cases, repositioning worked and sales increased.

An additional factor involved in positioning is called line-extension. Many firms fall into the line-extension trap simply because of poor planning. Sara Lee's entrées failed ($8 million loss), whereas its frozen baked goods are successful. Pierre Cardin wines failed, and so did Chanel's line for men. Bic pens are successful, but their pantyhose venture was a loss. Life Savers have become almost a generic candy, but the gum failed. The success of a company's product in the marketplace does not necessarily guarantee success for all its other products. This is particularly true if the firm has established itself firmly in one market with one product.

Some companies rarely line-extend. For instance, Proctor & Gamble products all receive a discrete identity; there is no strong corporate identification. Scott Paper Company does line-extend. Its "Scotties" name has been used for Scottissues, Scotkins, and Scottowels.

9. BRAND NAMES AND TRADEMARKS

Trademark. Any word, name, symbol, device, or any combination of these adopted and used by a manufacturer or MERCHANT to identify his goods and distinguish them from those of others. It is a BRAND NAME used on goods *moving in the channels of trade.* Rights in a trademark are acquired only by use, and ordinarily must continue if the rights are to be preserved. That provision is made to register a trademark in the Patent Office does not imply that such registration in itself creates or establishes any exclusive rights. However, registration is recognition by the government of the right of the owner to use the mark in commerce to distinguish his goods from those of others. BRAND is the everyday term; TRADEMARK is the legal counterpart. Trademarks are registered for twenty years and may be renewed every twenty years thereafter if not abandoned, cancelled, or surrendered.

Encyclopedia Brittanica

In the United States, the registration of trademarks began in 1870, when the Averill Chemical Paint Company registered its eagle, and when the William Underwood Company registered the red devil for its deviled-ham spread in 1876. Before this, trademarks were not a federal matter, but subject to common law, and a company's distinctive symbol could be defended in the courts. The fiercely competitive patent medicine industry made judicial history in this regard by giving rise to many court cases in the mid-19th century. Many of the trademarks used for patent medicines never changed, even though the formula, maker, and even advertisements changed with time. Radway's ministering angel and Lydia Pinkham's maternal countenance were known to several generations.

Although many in the advertising industry feel that a brand name should describe the product, be short, unique, easy to pronounce, and have graphic appeal, few names meet all these criteria. Mr. Clean, Green Giant, and Head and Shoulders are excellent examples of those that do. Names such as Easy-Off and Deep Clean are not especially unique, and they offer few graphic possibilities. And a name such as Men-E is not only difficult to pronounce but hardly descriptive of a shampoo. Pearl Drops is a good name for a toothpaste, but it comes in a container that looks more like nosedrops. Hardee's has a name that does not go very well with a juicy piece of meat. Walter P. Margulies, president of Lippincott and Margulies, packaging and communications consultants, recently noted in the *New York Times* that "the world is just running out of names. We have used computers to search through 128,000 words in order to produce one that was both appropriate and unclaimed."

One trend is the increased use of X and Z in deriving product names. First there was Xerox, then later, Exxon. Now there is Xoil, Xplor, Xicor, and Xylogics. X is the mark beside the dotted line, as well as the symbol for unknown quantities. Now it has a new function in the names of unknown corporate quantities.

Because of the paucity of meaningful names, executives have begun to invent words they hope will convey an alluring image and pique the curiosity of investors. "We like to see names that suggest a way-out futuristic product," says J. Morton Davis, president of D. H. Blair Company, a Wall Street brokerage firm, and founder of Xplor, a new oil exploration company. "Take a name like Xicor (pronounced Zycore), for instance. While it doesn't mean anything, it has a certain ring, a pizzazz and a fizzle to it."

For much the same reason, the Z has begun to spread, too, as in Zeus Energy and Zoe Products. Some companies are going even further afield. Gene M. Amdahl, the IBM émigré who founded the Amdahl Corporation, has founded a new company named

TABLE 9.1 TRADEMARK SYMBOL AND PSYCHOLOGICAL ROLE

Trademark Symbol	Product	Psychological Role
Father Knickerbocker	Beer	Mythical salesman
Betty Crocker	Foods	The "ideal" homemaker
Aunt Jemima	Pancakes	Goodness and abundance
Log Cabin	Syrup	Association with President Lincoln, thus engendering confidence
Life-Savers	Candy	Attention-provoking
Quaker man	Foods	Integrity and quality
Arm and hammer	Baking soda	Power (to leaven baked goods)
White Horse	Scotch	Purity, high ideals, power, victory
Dutch girl	Cleanser	Cleanliness, frugality
Camel	Cigarettes	Orient, Turkish tobacco, mystery
Borden's cow	Dairy products	Wholesomeness, purity
Bear (Hamm's)	Beer	Strength
Reddy Kilowatt	Electric utility companies	Lightning, instant illumination
Roadrunner	Plymouth automobile	Speed and reliability
Greyhound	Bus transportation	Speed and slickness

Acsys Ltd., which stands for Amdahl Computer Systems. Amdahl says that it has been well received by investors both here and in Europe.

Symbols created for brand names, such as animals, cartoons, people, birds, and other animate objects, include many examples, a few of which are listed in Table 9.1. As Dr. Ian Haldane of the British Market Research Bureau states, symbols "are attitude-inducing and hence behavior-inducing, because they are intimately related to the values and norms of the society in which, willy-nilly, we find ourselves." But sometimes a company chooses the wrong symbol. In the 1960s, Proctor & Gamble created the cartoon character, "Jifaroo," to promote "Jif" peanut butter, but it was later deemed inappropriate for a nutritional product. Such characters must also be periodically updated to conform with the style of the time. Borden's "Elsie" has gone through several appearance changes since her original artistic inception. There have been six "Betty Crockers" since 1936. And the "umbrella girl" of Morton Salt has also been modified over the years.

Advertising executive David Ogilvy calls the trademark "a first-class ticket through life" and designs one for each of his clients to raise their social status. Choice of brand names, packaging, texture of cartons, and typeface style each make their distinctive contributions to the success of a package. A brand image also invokes past association because the product becomes deeply ingrained in the consumer's mind over the years.

Brand names and trademarks have become an integral part of American life. When rumors circulated that there was no Mrs. Smith who made pies, Mrs. Smith's Frozen Foods Company, a subsidiary of Kellogg's, quickly promoted its pies with a TV commercial featuring Mrs. Smith's son, who founded the company. "I'm proof she was real," Mr. Smith says on the TV spot.

There are two stories about the origins of the Quaker man used by Quaker Oats. One is that it was chosen by Henry D. Seymour, one of the founders of the firm, as a means of connotating purity, quality, strength, and manliness. The other is that Seymour's partner, William Heston, got the idea from a picture of William Penn in Quaker garb. It may also have originated from patent medicine. Pitchmen wore Quaker costumes when they sold their remedies. These merry gentlemen, more thespians than doctors, caught the public fancy. One leading patent medicine of the day was Dr. Flint's Quaker Bitters, showing a Quaker man on the label exclaiming the virtues of the product.

AUNT JEMIMA

> Shrove Tuesday, Shrove Tuesday
> 'Fore Jack went to plow
> His mother made pancakes,
> She scarcely knew how.
> She tossed them, she turned them,
> She made them so black
> With soot from the chimney
> They poisoned poor Jack.
>
> —From *A Shrove Tuesday Pancake Feast*

Of all the familiar trademarks, Aunt Jemima is one of the most appealing and expressive. Her beaming face has created a legend and endowed a rather common product with an appealing warmth

Fig. 9.1 *A rare Quaker Oats package, since it does not feature the "Quaker Man" on its front (see Fig. 2). Courtesy, Landor Associates, San Francisco: Museum of Packaging Antiquities; photograph by Jeanne Riemen.*

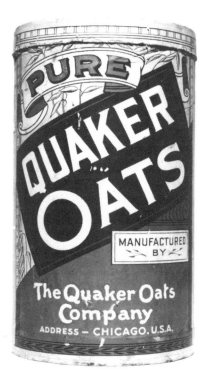

while establishing it firmly in the marketplace. She suggests abundance, pleasure, and happiness, and the consumer relates these to the product in the package.

The first pancake mix was formulated in 1888 by Chris L. Rutt, an editorial writer on the St. Joseph, Missouri, *Gazette,* and a friend in the milling business, Charles G. Underwood. After countless experiments, they developed a mix containing hard wheat flour, corn flour, phosphate of lime, soda, and salt, to which milk was added. One of the first taster's Purd B. Wright, later described it, "I ate the first perfected Aunt Jemima pancake and pronounced it good!" As reported in Arthur F. Marquette's book, *Brands, Trademarks and Good Will,* Purd's enthusiasm was so hearty that he was moved to break out a bottle of Missouri corn whiskey and share it with both Rutt and Underwood.

Rutt needed a trademark and package design that would reflect the festive spirit that had always been associated with the pancake. With the advent of voluntary Lenten shriving observances in 461 AD, the pancake became associated with the celebration preceding the holy season of Lent. In 1889 he established his place in American history by attending a local vaudeville house.

On the bill were the black-face comedians Baker and Farrell, whose show stopper was a jazzy New Orleans-style cakewalk to a tune called "Aunt Jemima," which Baker performed in the apron and red bandana of the traditional southern cook. "Old Aunt Jemima" was one of the most popular songs done by Billy Kersands, a famous black minstrel in the 1870s. Rutt chose "Aunt Jemima" for the new pancake mix because she was associated with good cooking. Rutt and his partner registered the trademark, but shortly thereafter they sold their company to R. T. Davis, owner of the Davis Milling Company. Davis decided to bring the trademark to life. In 1893 Davis launched a gigantic promotion at the World's Columbian Exposition in Chicago. His firm constructed a huge walk-in flour barrel that displayed and told the story of the new pancake mix. A former slave named Nancy Green was hired to play Aunt Jemima, and she quickly became the hit of the fair. More than 50,000 orders were placed for Aunt Jemima Pancake Mix, and at the end of the fair Nancy Green was awarded a medal.

By 1910 the name had become known in every state. Some competitors attempted to infringe on the trademark rights, but they were upheld by the courts in a suit in 1917. The name has not been seriously contested since.

The Aunt Jemima Mills were purchased by Quaker Oats in 1925. Over a period of eighty years, "Aunt Jemima" has become a national institution and a part of American folklore.

THE CREAM OF WHEAT CHEF

Forget the flour. Send us a car of "Cream of Wheat."
                    Hannah Campbell, *Why Did They Name It*

In 1890 George Bull and Emery Mapes of Grand Forks, North Dakota, acquired some machinery at a fire sale, and opened a small mill. Their head miller, Thomas Amidon, took home the "middlings" (farina) to be cooked with his breakfast cereal, and he persuaded Bull & Mapes to try selling some packaged middlings to grocery wholesalers. Mapes secured a small supply of cartons and for the label found an old woodcut of a black chef brandishing a skillet. The company adopted the brand name "Cream of Wheat" because the product was made with the whitest portion of the wheat. Mapes included ten cases of the new item with a carload of flour being shipped to a New York wholesaler. Instead of indignant complaints, the wholesaler sent a telegram for more of the same.

A few years later, in a Chicago restaurant, Mapes was impressed by his waiter's infectious grin and saw a substitute for the woodcut. The waiter was persuaded, for five dollars, to pose in a chef's cap for the famous full-face view that has continued to appear on millions of Cream of Wheat boxes.

IVORY SOAP

All thy garments smell of myrrh and cassia, out of the ivory palaces, whereby they have made thee glad.

                    Psalms 45:8

In 1878 experiments with a formula for a hard white soap without expensive olive oil were capped by a stroke of luck. The new mixture was poured into a blending machine, but apparently a workman forgot to shut off the power when he went to lunch. The result, a hard soap that floated! Customers who received the accidental batch called for more "floating soap." The company soon surmised that intensive whipping had beat air into the mixture.

Several weeks later, Harley Proctor, son of one of the founders of Proctor & Gamble, sat in church one Sunday morning struggling to follow his minister's sermon. He was thinking of this new soap that was white, pure, and floated. But the name he was using, "The White Soap," lacked force. Suddenly he became aware of the minister reading from the book of *Psalms* and heard the word *ivory*. The new soap was christened. Proctor then sought scientific backing for the claim of purity and sent samples to chemistry professors at Yale, Michigan, and Princeton for analyses. Their findings became one of the most famous slogans in advertising history: "99 and 44/100% pure."

The Bible has provided many other names for food and drug products. The original name of C.W. Post's corn flakes was Elijah's Manna, but this aroused the wrath of many influential ministers, and the British government refused to register the trademark, so Post changed the name to the now familiar Post Toasties.

Religious objections to product trademarks exist even today. The Underwood Devil on the firm's deviled-ham spreads was made into a "happy" devil in order to avoid religious problems.

In the history of patent medicine are products with names such as St. Anne and St. Joseph, Pastor Koenig, Father Francis, and the Good Samaritan. Nostrums were marketed bearing such names as Balm of Gilead, Paradise Oil, Resurrection Pills, and 666 (from *Revelation* 13:18).

Biblical names are used even today: "Samsonite" luggage, for example, and the Quaker man trademark of Quaker Oats, which was once defended against the Society of Friends, who petitioned Congress, unsuccessfully, to bar trademarks with religious connotations.

BORDEN'S "ELSIE THE COW"

> Calf: "Mama, I think I see a germ!"
> Cow: "Mercy, child, run quick for the Borden Inspector."
> From a 1936 Borden advertisement

In the 1930s the dairy industry had its share of public relations and consumer problems. Well-publicized "milk wars" raged between farmers and dairy processors. The big dairies were frequently pictured as evil moneymakers, taking advantage of both the farmers and the public. Borden concluded that the best approach was a friendly one.

In 1936 Borden's launched a series in medical journals that was to result in the creation of Elsie, the Borden Cow. The ads featured a variety of cartoon cows with names such as Mrs. Blossom, Bessie, Clara—and Elsie. A typical ad showed a group of young heifers hanging on the words of a lazy, unimpressive

Fig. 9.2   *Elsie, the Borden cow, and her son, Beau. Courtesy, Borden, Inc.; "Elsie" and "Elsie the Cow" are trademarks of Borden, Inc.*

cow. One heifer says, "And now tell us about the time you got kicked out by Borden's." Doctors loved the ads and requested reprints to distribute in their waiting rooms.

In 1938 Elsie came to life coast to coast in both the United States and Canada. Borden was then sponsoring a network news commentator named Rush Hughes. A radio copywriter, intrigued by one of the medical journal ads, prepared a commercial that so delighted Hughes he read it himself. It referred to the following letter:

> Dear Mama:
> I'm so excited I can hardly chew. We girls are sending our milk to Borden's now!
> > Love,
> > Elsie

Soon fan mail was pouring in, addressed not to Hughes, but to Elsie.

By 1939 Elsie had made her debut in national consumer magazines, and she and Rush Hughes continued their correspondence on the air. In 1939 Borden opened its scientific exhibit at the New York World's Fair. The focal point was a new *rotolactor,* a kind of futuristic merry-go-round where cows were automatically milked in a circle. Seven young hostesses answered questions and kept notes of the questions most often asked. At the first month's end the tally was: 20 percent about the rotolactor, 20 percent about the location of the rest rooms, and 60 percent about which of the 150 cows was Elsie.

With the growth of Elsie's popularity, Borden found that it would disappoint many people if it couldn't produce a real Elsie—and fast. Of all the cows in the exhibit, the most beautiful was a seven-year-old blueblooded Jersey from Brookfield, Massachusetts, whose name was "You'll Do, Lobelia." Borden renamed her "Elsie," dressed her in an embroidered green blanket, and put her all alone on the rotolactor twice a day for all to see. The public took her to their hearts, and Elsie became the spokescow for Borden ever after.

"PSYCHE," THE WHITE ROCK GIRL

> Psyche, the Goddess of Purity, has had an incredibly long run as White Rock's symbol. She's been sitting on the White Rock for more than a century.
>
> Venet Advertising can't claim we invented Psyche. But we did bring her to life on TV. When she ran through the forest, she brought new impact and new customers to White Rock. We've kept her as the focal point for all White Rock advertising ever since.—Venet Advertising, advertisement in *New Jersey Monthly* (1980)

One of the most famous trademarks of the Art Nouveau era is Psyche, the White Rock Girl, who has survived, with only a few overhauls, for more than eighty years. It started at the 1893 World's Fair in Chicago, when the owners of the White Rock Mineral Springs Company, a small firm bottling "Ozonate Lithis Water," discovered a painting by German artist, Paul Thurmann entitled *Psyche at Nature's Mirror*. It showed a solid Teutonic maiden, kneeling on a stone and peering at her reflection in a pool, minimally garbed and just sprouting the wings that denoted her transition from maiden to goddess.

The White Rock People bought rights to use the painting in advertising and inscribed "White Rock" on the stone on which Psyche knelt. In 1924 her hairstyle was changed to one then worn by Mary Pickford. In 1944 she was made taller and thinner and given an updated hairdo. In the 1970s her bra was removed.

MORTON SALT'S "UMBRELLA GIRL"

> When it rains it pours.—Anonymous proverb

The story behind the famous Morton Salt trademark began about 1910, when the Morton Salt Company was organized from the merger of several smaller firms. Most dealers were then purchasing 300-pound barrels of salt and dispensing smaller amounts to their retail customers. Efforts to get these dealers to buy three- or five-pound boxes of salt had met with minimal success. Morton's "Seal Salt," a high-grade table salt packed in a paper-lined bag, also had failed to meet with consumer approval. To again attempt to capture the imagination of the consumer, Joy Morton turned his attention to a new, free-running salt, which he packed in a spouted, round package. In 1912 Morton's Table Salt was launched in the blue-and-white, asphalt-laminated paper cannister with an aluminum pouring spout. This carton, invented by J. R. Harbeck, was eventually adopted as the standard for the entire salt industry.

The Morton Salt Company decided to advertise on a national scale, and in 1911 the little Umbrella Girl became a part of American commercial history.

The selection of the trademark was made by Sterling Morton II, then president of the company:

> One of the agency men suggested we might look at the three substitutes to see if we liked any of them better than the twelve which the agency considered best. I was immediately struck with one of the three. It showed a little girl standing in the rain with an umbrella over her head; under one arm she had a package of salt tilted backward with the spout open, and the salt running out. It struck me that here was the whole story in one picture. The message we wanted to put across—that the salt would run in damp weather— was made beautifully evident. I knew immediately that we could find no better trademark.
>
> Under the drawing of the little girl was the legend, "Even in rainy weather it flows freely." This struck me as being pretty good but rather on the long side. I distinctly remember saying that what we needed was something short and snappy like "Ivory Soap—It Floats." We worked around with "Flows freely, runs freely," but none seemed quite right. Finally, the word "pours" was suggested. That filled the bill, so "It Pours" as well as the words "Free Running" were approved for the new label.
>
> Then history was made. Someone (and I wish I knew who!) said, "There is an old proverb, It never rains but it pours." I think everyone in the room realized that we had something there. After a little discussion, I suggested that "never" and "but" struck me as poor words to use, that negative connotations should be avoided in a slogan, so we then turned the old proverb around and made it positive instead of negative—When It Rains It Pours. We knew that was it and our famous trademark and slogan were launched on their triumphant career.

PLAYER'S CIGARETTES' "BEARDED SAILOR"

> Player's Navy Cut cigarettes were launched in 1900 at 3d. for 10. By 1907 the brand was, next to Wills' Woodbine, Britain's largest selling cigarette.
>
> Chris Mullen, *Cigarette Pack Art*

One of the best-known trademarks in Great Britain for more than eighty years has been the bearded sailor on Player's Navy Cut cigarette package. It came as a surprise to many, in the summer of 1951, to learn that this was a portrait of a real sailor, Thomas Huntly Wood, who had served in the 1880s on the *HMS Edinburgh*.

Wood's likeness first appeared in the *Army and Navy Illustrated* in 1898, from which it was borrowed for advertising purposes, but the Player's sailor's head was registered in 1883. A friend of Wood's wrote to the firm in 1899 suggesting payment of £15; Wood reduced this to a sum of two guineas "and a bit of baccy for myself and the boys on board." The firm paid, in cash and product. Some time later, tired of people pulling out a pack of Player's and asking, "Is it really you?" Wood shaved off his beard. O. Henry or Damon Runyon could have built an enchanting story about Thomas Huntly Wood—the man who sold his face for a song, only to be haunted by it ever afterward.

UNEEDA BISCUIT'S "LITTLE BOY"

> I am Uneeda, I defy
> The roaming dust, the busy fly.
> For in my package, sealed and tight,
> My makers keep me pure and white.
>
> Anonymous (1900)

In 1899 the National Biscuit Company (now Nabisco) embarked on a packaging and promotional upheaval that *Fortune* magazine later claimed "did almost as much as the introduction of canned foods before it, and the invention of the electric refrigerator after it, to change the techniques of modern merchandising."

Fig. 9.3 *Gordon Stille, the "Uneeda Boy." Supplied by and reprinted with the permission of Nabisco, Inc., East Hanover, New Jersey.*

N.W. Ayer and Son advertising agency had coined the product name "Uneeda Biscuit" for the prepackaged soda crackers. A young copywriter named Joseph J. Geisinger thought of a fisherman, clad in a slicker, eating dry biscuits out of a dry package, which made the point that Nabisco products were always fresh and crisp. But the fisherman concept did not appeal to Adolphus W. Green, one of the founders of the company.

Geisinger asked his five-year-old nephew, Gordon Stille, a plump-cheeked, bright-eyed boy with a winsome look to pose in boots, oil hat, and slicker with a box of Uneeda Biscuits under his arm. Green approved, and the rest is history.

Also at the turn of the century, National Biscuit Company took its first important step toward creating its coat of arms. As the basic element, it used a symbol that has an ancient origin. In prehistoric times, the circle and cross with two bars represented the creation of life. In the early Christian era, it symbolized the triumph of the spiritual over the worldly. During the 15th century, this symbol was used as a printer's mark by the Society of Printers in Venice.

In 1900 Green suggested that it be incorporated in a coat of arms for the company. Since then it has undergone a number of changes, but the basic design and symbol remain much the same.

Today, the Nabisco coat of arms is a red triangular seal that spans the upper left-hand corner of all Nabisco packages.

GENERAL MILLS' BETTY CROCKER

How do you make a one-crust cherry pie?
What's a good recipe for apple dumplings?
How long do you knead bread dough?
                    —*Questions sent to Betty Crocker in the 1930s*

In 1921 the Wasburn Crosby Company, the forerunner of General Mills, used the name "Betty Crocker" in replies to homemakers' requests for recipes and help with baking problems. The surname was that of William G. Crocker, retiring director of the company's board. The name "Betty", was selected because it was popular at the time and was perceived to have a warm sound. In 1936 the first portrait of the corporate symbol was commissioned. The original Betty Crocker reflected the view of a 1930s housewife. She was a homemaker in her mid-40s and dressed simply and frugally.

The 1980 model is the sixth Betty Crocker. In her mid-to-late thirties, she is about ten years younger than her predecessor, who was introduced in 1972. But like the others before her, Betty Crocker VI is a blue-eyed brunette. Her portrait, like all Betty Crockers, graces the cover of the new Betty Crocker Cookbook, and it will hang, as do those of her predecessors, in a hallway of the Betty Crocker test kitchens at the company headquarters, where 75,000 visitors tour each year. But her image will not be featured on General Mills packages. The image of Betty Crocker III was the last to be used for that. Still, today 90 percent of the

Fig. 9.4   *The first Betty Crocker, 1936 (left) and her modern counterpart, 1980.*
Courtesy, *General Mills, Inc., Minneapolis.*

American public recognize Betty Crocker and relate her to General Mills.

Not everyone has responded warmly to Betty. In 1972 the National Organization for Women filed a class-action complaint charging General Mills with race and sex discrimination in perpetuating the image of Betty Crocker as a homemaker. Others, however, have been captivated by the Crocker image. Over the years, Betty Crocker has received hundreds of marriage proposals through the mail, apparently from men who believed she was real.

10 CONSUMERISM COMES OF AGE

Consumers are wary, and weary, of false claims, of advertising double-talk, of phony product comparisons, of overpromises, of rigged testimonials, of ersatz products, of shoddy quality, of being ripped off, of intolerable delays in delivery, and, in some instances, of total failure to fulfill.

Consumers are looking for, and demanding, honest value. They want and deserve credibility and trust. If consumers cannot trust us, they will do one of two things. They will desert us. Or, they will make sure that we are so bound up with restrictions and Government regulations that they can feel safe in our presence.

William E. Winter, keynote speech to
Association of National Advertisers Seminar (1980)

While consumer protection became an issue in the early 1960s, it was not a new idea. In England, consumer protection was noted as early as 1899 in *Mrs. Beeton's Book of Household Management*, the Victorian Woman's all-purpose home reference. It notes that "the act passed in 1872 for the prevention of Adulteration of Food, Drink and Drugs declares that persons who adulterate articles of food, or who sell those they know to have been adulterated whether with material injurious to the health or not, are punishable with fine or imprisonment." In addition, any Victorian consumer "could have any article of food, or drink, or drugs analysed by the public analyst of his district on payment of a sum not less than half-a-crown and not more than half-a-guinea."

PRODUCT SAFETY AND LABELING

In the United States, the seeds were planted in the early 1800s, and in the 1880s they began to take root, when members of Congress attempted to introduce a comprehensive national food and drug law. The first bill, introduced in 1879, and the Paddock Bill, passed by the Senate in 1892, specifically protected the medicine makers of the day from having to disclose their formulas.

In 1899 the Proprietary Association stated: "The demand for Pure Food laws is growing every year and, sooner or later, in all the States such laws are likely to be enacted." The first consumer organizations began to appear about this time. The Consumers

Fig. 10.1 *Kennedy's Medical Discovery. "Mr. Kennedy of Roxbury, has discovered in one of our common pasture weeds a remedy that cures Every Kind of Humor from the worst scrofula down to a common pimple." An excellent example of the kind of patent medicines sold before the 1906 Pure Food and Drug Act. Courtesy, collection of Edward Morrill, Werbin & Morrill, Inc., New York City.*

League of America, a labor-oriented organization, and the National Consumers League, founded in 1899, were among the earliest.

The tide shifted when Dr. Harvey Wiley campaigned for a National Pure Food Law. Wiley traveled around the country preaching that "when God's bounty was tampered with, the label should at least say so." But legislative action came slowly. Upton Sinclair's *The Jungle* changed the situation. Sinclair revealed the conditions under which America's meat was processed, how inspectors blinked while tubercular carcasses were brought back into the line, how rats and the poisoned bread put out to catch them were ground up with the meat, how employees now and then slipped into steamy vats and next went forth into the world as Durham's Pure Leaf Lard. The public was shocked, and meat sales dropped. Sinclair's revelations, President Roosevelt's agitations, and Dr. Wiley's initial zeal led to the 1906 Pure Food and Drug Act.

Control over "the purity and honesty of the food and medicines of the people are guaranteed," said the *New York Times* on July 1, 1906. The muckraking journalists and the era of Progressivism heightened consumer consciousness, and reformers began to talk more and more about the "consumer" and the "common man."

The period between the Progressive era and the New Deal saw few consumer protection crusades or legislation. The major campaign in these years centered on the standardization of containers and quality grading of foods. In the years prior to World War I, several acts were passed regulating and standardizing the size of some containers. But containers and packaging were not the concern of the reformers. Their main aim was to establish national standards of quality for food. The passage of the McNary-Napes amendment in 1930 helped; but not until 1938, when the "elixer sulfanilamide" tragedy occurred, was the last roadblock removed for the enactment of an improved 1906 Act.

The "elixer," a sulfa drug with diethylene glycol as a solvent, was never tested for safety, so the manufacturer did not know that the solvent was toxic until it killed 170 persons. The FDA could act only after the fact and only because of a technicality. The only violation of federal law was that the drug was mis-

Fig. 10.2 *The issue of child-resistant packaging is featured on these Dutch stamps, 1978. Courtesy, Minkus Publications, Inc., New York City.*

labeled an "elixer." The public pressure generated resulted in the Federal Food, Drug and Cosmetic Act, signed into law on June 25, 1938.

The new law was a tremendous improvement over the original one of 1906. It set the standard for all subsequent legislation in consumer protection, not only in the United States, but abroad. The 1938 law did the following things:

1. Prohibited foods dangerous to health
2. Prohibited unsanitary packages
3. Established the Food and Drug Administration
4. Plugged many loopholes in the 1906 law
5. Established stiffer fines for offenders
6. Enabled courts to issue injunctions against repeated violations

From the days of the New Deal until the early 1960s, consumer protection was part of a cluster of major social issues rather than an issue in itself. Perhaps the only exception was the Food Additives Amendment of 1958. The sparkplug for this legislation started in 1956 at a cancer seminar in Milan, Italy. The doctors theorized that one cause of the world increase in cancer was the increased use of chemical additives in foods. This news was acted upon by Representative James J. Delaney, a Democrat from New York, and Senator Estes Kefauver, a Democrat from Tennessee, who moved to ban cancerous substances from food. The result was the Food Additives Amendment, which defined an additive as any substance that directly or indirectly becomes a part of or affects the characteristics of food. It shifted the burden of proof from government to industry in establishing the safety of an additive and established as law the Delaney Clause, which set a zero tolerance for any additive that is a known carcinogen, specifying that "no additive shall be deemed to be safe . . . if it is found, after tests which are appropriate for the evaluation of the safety of food additives, to induce cancer in man or animal."

Passed in a time when testing procedures and tools were not as sophisticated as today, the amendment means little in 1980. In packaging, this amendment has been responsible for the demise of the rigid PVC bottle for alcoholic beverages, the suspicion about PVC in all areas of food packaging, and the ban on acrylonitrile-based beverage packages.

There has been no movement to change the amendment or repeal it, and it could easily hamper the development of many packaging materials.

Congressional hearings led by Senator Kefauver in 1962 led to further amendments. In 1965 the National Bureau of Standards, working with an ad hoc industry group, developed a model state weights and measures law, and in 1966 the Fair Packaging and Labeling Act was passed. Introduced in 1962, by Senator Philip Hart from Michigan, the bill was to facilitate price and quantity comparison of retail grocery items by shoppers; in short, it introduced price competition in the retail outlet by standardizing package sizes to allow price comparison. The final version of the bill allowed for voluntary limitation of a number of package sizes.

Following close on the heels of the FPLA was the cosmetic-ingredient labeling requirement, a mandatory nutritional labeling requirement for any food that makes a health claim.

In the 1970s consumer activists in the Senate, such as William

Proxmire (D-Wisc.), Abraham Ribicoff (D-Conn.) and Joseph Montoya (D-N.M.), and in the House, such as Thomas Foley (D-Wash.), Benjamin Rosenthal (D-N.Y.) and Neil Smith (D-Iowa), worked actively on behalf of consumer protection.

The most significant legislation to emerge was the Consumer Product Safety Act of 1972, which coordinated the administration of the Flammable Fabrics Act (1953), the Federal Hazardous Substances Act (1960), the Refrigerator Safety Act (1969), and the Poison Prevention Packaging Act (1970).

Its administrative body, the Consumer Products Safety Commission (Address: Office of the Secretary, 1111 18th Street, N.W., Washington, D.C. 20207), was made up of five commissioners appointed for seven-year terms by the President; one person was designated commissioner. Only three members could come from the same political party, and there could be no conflict of interest. The CPSC soon established the Product Safety Advisory Council, made up of fifteen members—five from industry, five from government, and five active consumerists—who could stipulate performance, composition, contents, design, construction, finish, or packaging of products. In 1981 the Reagan administration considered dissolving the CPSC, but did not.

Other programs instituted by consumer advocates include:

1. Aerosol ban—outlawing of fluorocarbon aerosol packaging
2. Rooney Bill (from UNWRAP)—EPA given authority to promulgate regulations for products which might use an unreasonable amount of energy or virgin material
3. CPSC legislation—controls on potential hazards of packaging carbonated drinks in glass bottles
4. Alcoholic beverage labeling—ingredient labeling
5. State net-weight regulations—proliferation of state laws that affect a nationally marketed product
6. Minnesota packaging restriction—authority to ban sales of package if judged inconsistent with state policies
7. Unit pricing—display price per unit weight or count
8. Open dating—coding each package of food with suspend date
9. Additive labeling—prominently marking all packaged foods containing additives

10. Imitation labeling—a New York statute barring "artificially made" food products from posing as equal to "natural" products

One of the more interesting laws under the jurisdiction of the CPSC is the Poison Prevention Package Act (PPPA) of 1970. Since the passage of the first regulations, those governing packaging of aspirin (1973), accidental ingestion by preschool children has decreased 41 percent, and aspirin-induced deaths are down 63 percent. Specially designed child-resistant closures and packages have contributed to the success of the PPPA. The test procedure to determine whether packaging is child-resistant and adult-effective is specified in the regulations. Under the regulation, 90 of 100 adults tested must be able to open and properly resecure a package, given the instructions that appear on it. The method to determine properly resecured packages is not described. In the past, both visual inspection by the tester and machine torque testing were used. Recently, CPSC has informed companies that products have failed under a third test used by the commission. Children were given bottles that adults had previously opened and closed. The commission found that children opened these bottles more easily than factory-closed bottles. Such bottles were judged neither adult-effective nor child-resistant. In 1978 Miles Laboratories challenged the regulations in court. A consent decree resulted in the return of all "One-A-Day Plus Iron" vitamin stock in the manufacturer's warehouses to be repackaged in containers more readily secured by adults.

WEIGHTS AND MEASURES

Control of packaged goods is not new. Egyptian pictographs show government officials weighing and marking bales of goods. Arab merchants in the 8th century used measuring cups and bowls certified by a bureau of standards and marked as to their contents. As early as 1266, in England, a law called *Statutum de Pistoribus* regulated the size of bread loaves, and a later regulation required bakers to stamp their mark in each loaf to certify compliance with the law.

Despite these efforts, weights and measures in the Middle Ages varied between one district and another. Some purchases were not measured or weighed at all. There is frequent reference to buying such items as "a salmon as thick as a man's arm," as much hay "as a man can lift," as much wood "as a man may bear," and cloth "two ells wide between the fists." English drapers bought cloth by the "yard and a hand." Coal was sold by the "seam," "load," "corf," "fother," "keep" and "chauldron."

Until about 1300, standard measures often were fixed by declaring a particular vessel to be the standard; then copies of the vessel could be made and purchased. Scales were expensive and kept by only a few merchants. Consumers bought everything by appearance. According to the Butcher Company's regulations in 1607, lamb must not be cut "deceitfully," but must have ten ribs in the forequarter and three in the hindquarter, and no piece cut off in between could be sold. And as late as 1667, a pamphlet entitled "What England Wants" urged that most food exposed for sale in markets and shops should be sold by weight, "as is done in Spain."

Control of weights and measures in the United States has historically (but not exclusively) been limited to the individual states.

Coordination of state weights and measure efforts, with attempts to create uniformity in the separate state regulations, is maintained by the Federal Bureau of Weights and Measures and the quasi-public Association of Food and Drug Officials of the U.S. Although model laws have been drafted, great variations in state regulations still exist.

New Jersey is an excellent example of a state with fairly tight control. Its jurisdiction goes far beyond packaging weights and measures: It also controls all types of retail liquid measuring devices and solid-weight scales, from gas pumps and lubricant dispensers to retail grocery and pharmaceutical scales. The state also subscribes to the model law created by the National Bureau of Weights and Measures.

CONSUMERISM VS. PACKAGING

The packaging industry has received more than its share of criticism from the consumer, and perhaps now it faces its greatest challenge ever. New criticisms against the industry spring up regularly and if these attacks on packaging continue unabated, the price of all packages could very well increase far beyond the cost of the product inside. The paradox of these efforts is that increasing governmental control on packaging may lead to exactly the opposite results of the requirements of the consumer—economy and uniformity.

Dr. Aaron L. Brody, in his article "Impact of External Influences on Food Packaging" (*CRC Critical Reviews in Food Science and Nutrition,* September 1977, pp. 227-64) states, "When packaging is threatened, so then is our food supply and way of life. If the pace of restrictions being threatened or implemented from sources external to the free market economy is permitted to continue, American consumers will pay disproportionately higher prices for food, our quality of life will be undermined, and, ultimately, hunger could result in a significant portion of our population. . . . Regardless of who has paid the direct cost to date, the consumer pays the ultimate bill."

Nutritional Labeling

Three major programs from the 1960s—nutritional guidelines, fortification guidelines, and nutritional labeling—are now being finalized and will be instituted on a voluntary and/or mandatory basis in the 1980s. Although some of these programs are totally or in part voluntary, governmental agencies and consumer actions could make all of them mandatory. Such issues are of direct concern to all members of the packaging industry.

Under a regulation enacted in 1972, which became effective in 1975, food packages that bear *any* statement concerning the nutritional value of the product must have a uniform and accurate description of its nutritional content. A 1978 study conducted by the Office of Education reported that less than 38 percent of American adults can comprehend what RDA (Recommended

Daily Allowance) and other nutrient information means on a package label.

The cost involved in instituting a nutritional labeling program can include:

1. *Initial Analytical Tests* ($500 to $5,000 per item)
2. *Label Design Changes* ($2,000 to $60,000 per product)
3. *Total Initial Costs* (up to $1 million)

These estimates do not include the cost of additional research that might be necessary to reformulate the product. A study by the Grocery Manufacturers of America showed that depending on the size of the company, average initial costs were approximately $18,000 for a small firm to almost $2.5 million for a firm with sales in excess of $1 billion.

Good Manufacturing Practices (GMPs)

Government has also taken a close look at how food processors operate. They are continually upgrading their efforts in food plant inspections. Governmental regulation of food plants is aimed at ensuring the safety of packaged foods. GMP was initially instituted because of the failure of several food processors to adhere to recommended technical guidelines proposed by the industry. GMPs for the food industry deal with recommended practices for handling and processing of foods as well as guidelines for packaging.

Because of the new GMPs, converters of packaging materials have become overly cautious about the material shipped, and food processors are skeptical about introducing new products. This added and often unnecessary attention costs money, and these costs are being passed on to an already overburdened consumer. Under the present Administration, a loosening of some of these unnecessary regulations is anticipated.

Solid Waste Disposal and Litter

Consumer concern about the effect of packaging on solid-waste disposal has become a matter of substantial interest. Packaging constitutes less than 13 percent of all solid waste and less than half of municipal waste in the United States. Yet consumerists argue that banning high-volume, nonreturnable beverage packaging would have a significant, favorable impact on the solid waste problem. To this end, more than 1,500 measures have been introduced in all states between 1969 and 1980 in an effort to force returnables by banning, taxing, or requiring a large deposit on one-way containers.

The solutions to solid-waste problems revolve around the application of advanced methods of waste management and technology. More efficient disposal programs are needed, coupled with the further extension of recovery and recycling technology.

A 1970 study conducted under the guidance of *The Litter Control Letter,* a publication of the National Council of State Garden Clubs, and underwritten by the Glass Packaging Institute and the National Soft Drink Association, identified 150 items in significant accumulations at high-litter locations. The study identified urban locations as suffering the greatest concentration of discarded items. The sixteen most prevalent included packaging wrappers and containers, building materials, tires, and newspapers.

Industry's most practical approach to the litter problem is to provide financial support for agencies and programs whose efforts are directed toward the prevention or elimination of littering and litter.

Packaging personnel believe that the greatest threat to the industry will be in the area of future state and local regulations. Fully informed, intelligent and proper regulations are not a threat. The difficulty lies in the passage of well-meaning but foolish regulations that could effectively destroy an industry and indirectly affect a nation's economy.

Future Federal legislation appears to be limited. Although the Carter Administration was favorable to restrictive packaging legislation, the Reagan Administration believes in less restrictive legislation. It is anticipated that the next four years will see an enhanced free-flow of packaging.

11 PACKAGING FOR MILITARY AND FOOD-SERVICE MARKETS

Do the G.I.s still complain about the food? Of course. They would complain if they were cosseted by the finest cooks in the world. They will continue to complain, swearing that the mess sergeants are venal, that the best food goes to the officers, that the beef was rejected by General Sherman. They will also continue to eat.

Drew Middleton, *New York Times* (September 5, 1979)

There are two markets served by packaging that require almost no sales appeal or advertising. Instead, the military and food-service markets emphasize product protection, identification, marking, and physical handling while military packaging stresses protection at the sacrifice of convenience, food service packaging demands convenience at the sacrifice of long shelf-life.

THE MILITARY MARKET

Many significant packaging innovations have originated in war-time and from the requirements and demands of the military.

Early examples include Caesar's grain trains, which supplied his troops with their staple, boiled wheat, and needed protection against the twin ravages of pests and cold. Probably still more significant was Nicholas Appert's discovery that moist heat could be used to destroy microorganisms in food, a direct result of Napoleon's requirement in feeding his army. Instant coffee made its first appearance among the doughboys of World War I, and Spam went on to sell well in the civilian market after World War II.

Many other packaging developments came out of World War II. During the war, thermoforming was used to make contour maps and later, in the 1950s, the technique came into general use. Protective packaging was developed for dehydrated foods. The onset of World War II catalyzed significant American developments in food technology. Soldiers needed lightweight foods that could be transported easily in war zones. Thus an entire range of food rations were developed.

The modern packaging of Army rations began during the Boer War (1900). Most American forces in World War II subsisted largely on K-, D-, and C-rations. K-rations consisted of breakfast, lunch, and dinner cans packed in a wax-dipped cracker-box-size unit. However, they lacked variety and essentially were concentrated. C-rations, also canned, and containing about 3,600 calories for a day's supply, were introduced at the end of World War II and lasted until 1959. D-rations, enriched candy bars packed in a flexible overwrap in a paperboard carton, were emergency rations and not used after 1944. They were very rich, and many GIs complained of nausea after eating them. British troops (especially the British First Army) had Compo rations, consisting of enough canned food for fourteen men for one day. The cans contained such hearty dishes as steak and kidney pie and plum pudding. Once introduced to "Compo," many Americans opted for the British ration. Trades were common, with one enterprising British parachute officer trading ten cases of "Compo" for a Jeep.

World War II also brought the "easy-open" can, a "tin" with an attached "tin opener." Also the first glimpses of medical packaging began to appear during and shortly after the war with plasma

Fig. 11.1 *K-Ration carton.* Courtesy, *Landor Associates, San Francisco: Museum of Packaging Antiquities.*

packaging and various types of pharmaceutical packages. Streptomycin, penicillin, and sulfa drugs became standard prescriptions as a result.

After the war, the rise in consumer affluence coupled with the demand for convenience created an interest in dehydrated foods. Many reconstituted, "mock" foods were produced: dehydrated milk and eggs, dried potatoes, whole meat, Spam, and in England, a curious fish called "snoek." Canned foods were rationed because tin went into armaments and cans for soldiers' K- and C-rations.

Several manufacturers attempted to persuade Washington that their product was essential to the war effort and thus should not be subject to wartime rationing. One classic example was Wrigley's chewing gum. Chicle, like rubber latex, came from trees in Japanese-occupied Southeast Asia. But Wrigley's could obtain South American chicle easily. The company had its gumtree tappers tap the rubber trees growing in the same area. Both the chicle and the high-priority rubber were transported back to the United States on the same ships. Next, to obtain a sugar priority from Washington, Wrigley argued that chewing gum relieved wartime tensions. The net result was that Wrigley's began to supply a stick of gum for each package of K-rations.

There were also several packagers that converted from their conventional product to those needed for the wartime economy. A soft-drink producer, accustomed to filling bottles with liquids,

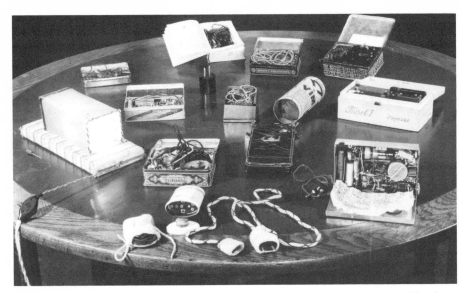

Fig. 11.2 *These tins and other packages held secret receivers and were used by the Dutch underground in World War II. Courtesy, Philips Persdienst, Eindhoven, The Netherlands; photograph, Fotostudio Peter Voorhuis.*

instead loaded shell casings with explosive chemicals. A paper-box manufacturer turned out plasma-kit containers. A bedspread manufacturer converted to mosquito netting. There were also many shortages of packaging materials, particularly tinplate. Both the manufacturers of Times Square and Home Comfort tobacco switched from their conventional tinplate package to cardboard containers. Both the tin and cardboard containers are collector's items today. Even the ordinary paper bag was modified. Bags used in grocery stores were often made longer so that they could hold larger quantities of products for the homeward-bound shopper.

Packaging Requirements

All products intended for the Armed Forces must be delivered in a usable condition. Since the ultimate destination of the packages cannot always be anticipated at the procurement stage, efficient packaging is the only solution, and the packaging must be of a higher quality than would be needed commercially. See Table 11.1 for some differences between military and commercial packaging.

The two major risks facing packaged items for military use are physical injury during shipment and deterioration as the result of climate and unusual storage conditions. Some storage areas might be subject to excessive humidity, salt air, even low-level radiation. Military equipment is often procured years before it will be used, and this creates special inventory management requirements. In many cases, all spare parts for the life of the equipment must be procured at the same time and stored with the parent equipment. Thus their packaging must ensure protection over a long period.

Fig. 11.3   *To stay in business during World War II, Philip K. Wrigley had to demonstrate that his chewing gum was essential to the war effort in order to obtain a sugar priority. He argued—with the help of a government laboratory—that chewing gum was a great reliever of wartime tension. He wound up not only supplying a stick of gum for each package of Army K rations, but also packing the rations in his factory. Courtesy, Food Packaging and Processing Group, Food Technology Division, Food Engineering Laboratory, Department of the U.S. Army.*

TABLE 11.1 COMPARISON OF MILITARY AND COMMERCIAL PACKAGING REQUIREMENTS

Factor	Military Requirement	Commercial Requirement
Protection	When item is unpackaged, it must be able to perform its designed function; protection must achieve 100-percent reliability.	Certain percentage of damaged items usually tolerated. Amount depends on factors such as cost of item, profit, and insurance cost.
Logistics	Destination and means of transport usually unknown.	Destination and means of transport usually known.
Storage	Duration usually unknown; long durations possible.	Type and duration generally not severe and usually known.
Cost	Minimum consistent with protection requirement.	Minimum consistent with cost of item, profit factor, tolerated damage, and use of package as sales vehicle.
Evaluation of package	Difficult to assess because of length of storage.	Immediate feedback from wholesalers and retailers.
Methods and materials	Must meet specifications and standards.	No limitations.
Package weight, cube, and configuration	Minimum weight and cube. Best configuration for ease of handling and minimum cube.	Often determined by merchandising conditions, such as type of display racks and appeal.

Source: *Australian Packaging*, August 1980.

All packaging of military stores must be designed with the end user of each item in mind, and it must be able to protect the items in quantities. For example, it is of no value to provide a perfectly preserved case of 100 wheel bearings for use in a field repair situation if the rest of the case would deteriorate as soon as a small quantity is removed.

Another factor to be considered in military packaging is standardization. An item may be produced by many different firms and adequately controlled specifications are needed to ensure that the product's quality is both uniform and exact. This standardization of materials and methods ensures that the package will protect its contents until required for use. Service packaging is done by manufacturers of the items, contract packagers, specialist service units, supply units, and user units in base installations or field areas. In storage installations, packaging is inspected regularly to see that it has not deteriorated.

Because military packaging demands a higher degree of preservation, identification, and physical protection than its commercial counterpart, many packaging firms are cautious about accepting a military contract. They feel they would then be subject to countless inspections, specifications and regulations. For a small firm, a military contract often proves to be an excellent business venture; however, for larger firms, the hassle involved in serving the military is often not worth all the effort involved.

THE FOOD-SERVICE MARKET

The term *food service* refers to all activities related to the serving of food *outside* the home. Food service is the fourth largest industry in the United States in terms of sales, and it employs the largest number of workers. In 1979, the 400 largest food-service organizations in the United States achieved sales of $55.8 billion. The top 50 are:

1. McDonald's
2. KFC
3. Pillsbury
4. U.S. Dept. of Agriculture Food and Nutrition Services

5. Marriott
6. Holiday Inns
7. PepsiCo
8. ARA Services
9. Wendy's International
10. International Dairy Queen
11. U.S. Army
12. Hardee's Food Systems
13. U.S. Navy Food Service Systems Office
14. Denny's
15. Saga
16. Canteen
17. Sheraton
18. Howard Johnson
19. Army and Air Force Exchange Service
20. Foodmaker
21. Sambo's Restaurants
22. General Mills Restaurant Group
23. Ponderosa System
24. Best Western
25. Arby's
26. Jerrico
27. Servemation
28. Collins Foods International
29. Church's Fried Chicken
30. Morrison
31. Shoney's
32. Bonanza International
33. Hilton International
34. Tastee Freez International
35. Ramada Inns
36. Wienerwald U.S.A.
37. Burger Chef Systems
38. Hilton Hotels
39. Sonic Industries
40. Western International Hotels
41. Gino's
42. U.S. Air Force

43. Interstate United
44. W.R. Grace
45. Dunkin' Donuts of America
46. Baskin-Robbins Ice Cream
47. Hyatt Hotels
48. Intercontinental Hotels
49. Chart House
50. K mart

McDonald's hit the $5 billion mark, and eight other companies reported sales of $1 billion or more. When the other sectors of the food-service industry are added to this total, overall sales for the entire industry is well over $500 billion. In-flight food sales alone came to $1 billion in 1980, despite a slump in passenger traffic. The major in-flight catering companies are:

1. Marriott In-Flite Services
2. Dobb's Houses
3. United Airlines, Food Service Division
4. Sky Chefs (a subsidiary of American Airlines)
5. Ogden Food Travel Services
6. Air La Carte
7. Host International

While the food-processing industry has taken advantage of technological advances to increase its productivity and reduce its costs, the food-service industry has continued to use the methods employed since its earliest days. Until fairly recently, each food-service operation remained an on-site food factory and retail sales establishment. While other industries adopted the concepts of mass production, the food-service industry did not undergo its own industrial revolution until after World War II. Even today the vast majority of food-service organizations continue to produce the food from raw ingredients and sell the finished product at the same location.

As new methods of food service were developed to take advantage of the many technical advances in packaging, it soon became evident that on-site food production could be reduced or eliminated only through the use of partially or fully prepared foods—

canned, dried, chilled, or frozen. It also became evident that food packaging and the equipment for handling and preparing convenience foods had to be planned on a systems basis because food, packaging, and equipment procurement are closely related. The entire system must be fully integrated and planned properly to take advantage of new technology.

FOOD-SERVICE BUYING FACTORS

The operator of a food-service establishment is influenced in his buying habits by certain basic factors. Among the more important are:

1. The style of service, including the menu
2. The type of customer being served
3. The size of the operation
4. The location of the operation
5. The structure of the operation

As an extreme example, it would be foolish to try to sell select filet mignon to the operator of the corner hamburger stand. Perhaps equally important in determining buying habits are the characteristics of the customers, such as age, health, financial status, and ethnic background.

Of course, the size of an operation will influence the volume of buying, but it will also have an effect on the buying habits of the operator. For instance, a large modern operation, such as a well-equipped hospital, may have more adequate freezer facilities for long-term frozen food storage.

Location may influence the type of service that an operator expects from his suppliers. In major cities, he will expect prompt service from distributors, and he may also want more tightly competitive pricing. In more remote areas, where competition may not be so strong, an operator may be willing to accept somewhat slower service because of shipping delays.

And finally, the buyer for a highly centralized, multiple-location organization will operate differently from a buyer who purchases for a loosely knit organization, even though both

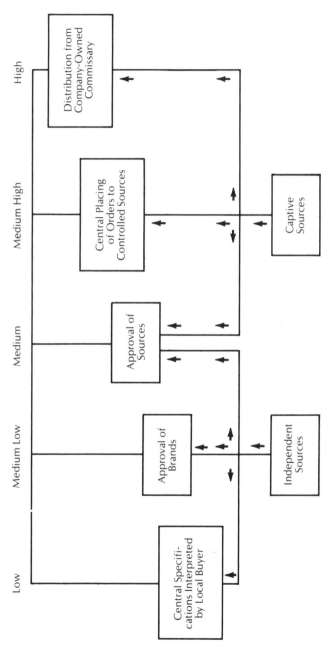

Fig. 11.4 *Degree of central control over purchasing.*

buyers may control the same amount of buying power (see Fig. 11.4).

Operations might use aluminum-foil containers, paperboard, plastic trays, plastic films, or aluminum-foil overwraps. By its very nature, food service packaging is meant to be simple and as standard as any package can be. The packages are rapidly discarded and are not intended for long shelf-life items. Food service packaging is generally purchased through a standard catalogue composed of basic versatile types of packaging.

Non-stock items are also supplied to the food service industry. The quantities needed are carefully considered before the packaging supplier offers these products to the market.

As an example of the types of packaging materials used by food service operators, a typical supplier of materials might have in stock:

1. Bulk rolls of foil, paper, plastic
2. Convenience rolls of the above, in shorter, multi-pack rolls
3. Sheets of foil, paper
4. Laminates of the above, such as foil/paper (used as sandwich wraps)
5. Bags (to hold products such as ice cream, hot dogs, etc.)
6. Foil and plastic to wrap around food containers
7. Cartons, all varieties
8. Containers and lids

The food-service market can be divided into four categories:

1. Commercial feeding
      Fast-food operations (most)
      Restaurants and cafeterias
      Caterers
2. Semicommercial feeding
      Schools and colleges
      Clubs, including service clubs
      Airlines and other transportation
3. Institutional and charitable feeding
      Hospitals and nursing homes
      Prisons and jails
      Convents and seminaries

4. Government feeding
      Military personnel
      Civilian personnel

The Commercial Feeding Market

More than three-quarters of all food-service sales come from the commercial feeding market. Since the customer in a commercial eating place pays for the food he orders, his satisfaction is of primary importance to the commercial food-service operator. Commercial eating places range from the hot-dog stand in an amusement park to the luxurious dining room (a "white table-cloth" operation) with candlelit tables and a French maitre d'hôtel in formal attire. Somewhere in between these two extremes is the typical American fast-food establishment, where the public spends some 80 percent of the money that goes to commercial food-service operators.

Fast-Food Operations

International Foodservice Manufacturers Association (IFMA) defines a fast food as a menu item that can be "ordered, prepared, dished up, served, and consumed within 30 minutes." A strict definition of fast foods implies that they are consumed on the premises, but most carry-out operations are included in the fast-food category. For example, Kentucky Fried Chicken considers itself a fast-food operation, even though some 80 percent of its orders during cold weather are taken home.

Because a large percentage of the fast-food operations in the United States are chain-owned, operated either by the chains directly or through franchises, much fast-food buying is done through centralized purchasing centers in which varying degrees of control exist.

Depending upon the chain (whether corporate or franchise), and upon its corporate purchasing policy, headquarters involvement in purchasing may vary all the way from merely setting quality standards to the operation of a central commissary. In a corporate operation, individual operators purchase everything they use. In a franchise operation, they purchase nothing themselves, although they are charged for the items they order from the commissary.

Restaurants and Cafeterias

A small corner restaurant is far removed from a luxury-hotel dining room, and both seem to have little in common with a downtown supper club. But both are in business to make a profit, of course, and both must please their customers.

Luxury dining rooms, whether they operate independently or in connection with a hotel, usually specialize in two commodities: gourmet food and high-quality service. As one IFMA publication states: "Elaborate dining is the primary entertainment in a luxury restaurant. . . . The real show is put on by professional service personnel who create an aura of . . . extravagance."

Suprisingly, some luxury dining establishments feature take-out service as a regular part of their operation. This is smart business because it adds income without a substantial increase in overhead costs. Although not all hotel or motel restaurants offer the service, quality, and price of the luxury dining room, the operations are similar. As part of the whole establishment, food service may be considered incidental to lodging. Indeed, in many hotels and motels, the food service operation is separate from the lodging function, perhaps even having a different ownership.

Hotel and motel food-service operations can offer different types of service, including table, counter or coffee shop, buffet or smorgasbord, catering, room, and vending-machine service. Larger units may have bars or lounges that may or may not offer food in addition to beverages. Obviously, the more services available, the more involved the food-service operation will be, and the greater the opportunities for the alert food-service salesperson. But the buying function may be more involved too, so it should be investigated and analyzed before a sales call is made.

Caterers

Catering facilities that use food service techniques may range from large banquet halls to smaller operations that can cater for a minimum of 5 to 10 people. Many caterers will prepare the food in their own kitchen and then either transport it pre-portioned or in bulk to the diner. In these situations, the packaging must be able to keep the food hot and appetizing. Because of the large quantities prepared, steam-table-size aluminum trays are often used. There are caterers that only prepare certain foods while others are more versatile and do offer a wider variety of products.

The Semicommercial Feeding Market

Another major market in the food-service industry is one where the food service is of secondary importance to some other activity. In fact, the food-service function in such operations owes its existence to the other activity. The profit motive may not be as strong as in commercial feeding, although customer satisfaction is still an influence. In the semicommercial feeding market, the greatest concern is with the best buy for the dollar spent.

Schools and Colleges

One important segment of the semicommercial feeding market involves the wide range of educational facilities, from elementary schools to universities. These obviously meet the criteria of semicommercial operations, since feeding is a function secondary to the primary activity—education.

Although grouped here as one segment, educational facilities must be divided into two subgroups because of the National School-Lunch Program, which exerts considerable influence on the buying practices of elementary and high schools, both public and nonprofit private. (The other subgroup consists of colleges and universities, plus those private schools that are, at least nominally, operated for a profit.)

The National School-Lunch Program provides federal financial assistance to those schools (or school districts) whose lunches meet certain minimum standards for nutrition and content. (The Type A lunch under the Program must contain one-half pint of fluid whole milk, specified amounts of vegetables or fruit, and certain combinations of protein foods like meat, fish, or eggs (see Table 11.2). The assistance may be in the form of outright pur-

TABLE 11.2 TYPE A SCHOOL-LUNCH GUIDE

Pattern	Pre-School Children (3 to 6 Years)	Elementary School Children		Secondary School Children (12 to 18 Years)
		6 to 10 Years	10 to 12 Years	
Meat and/or alternate (one of the following or combinations to give equivalent quantities):				
Meat, poultry, fish	1½ ounces	2 ounces	2 ounces	3 ounces
Cheese	1½ ounces	2 ounces	2 ounces	3 ounces
Egg	1	1	1	1
Cooked dry beans or peas	¼ cup	⅓ cup	½ cup	¾ to 1¼ cups
Peanut butter	2 tablespoons	3 tablespoons	4 tablespoons	4 to 5 tablespoons
Vegetable and/or fruit	½ cup	¾ cup	¾ cup	1 to 1½ cups
Bread	½ slice	1 slice	1 slice	1 to 3 slices
Butter or fortified margarine	½ teaspoon	1 teaspoon	1 teaspoon	1 to 2 teaspoons
Milk	¾ cup	½ pint	½ pint	½ pint

chase for a school district of such commodities as meat, orange juice, or flour. Alternately, the program may supply money to a school district to subsidize the cost of lunches prepared for children enrolled in the schools in that district.

Because of the National School-Lunch Program, which is in effect in about half of the nation's elementary and high schools, the United States Department of Agriculture (USDA) is by far the largest purchaser in the semicommercial food-service market. The USDA purchased well over $1 billion worth of commodities and supplies in 1974, serving 4.1 billion meals to 23.5 million children. In addition, several large cities invested heavily in food service for school children, including New York City ($128.8 million), Los Angeles ($79 million), Chicago ($65.5 million), and Miami ($36.1 million).

Elementary and high schools differ considerably from colleges and universities. In contrast to the "captive" patrons of elementary-school food service, college and universities must cater to the tastes and desires of young adults. Such giants as the University of Wisconsin ($25.8 million annual food-service sales) even have professional live entertainment in their snack food/ beer service areas. Many college food-service operations (and even some in secondary school systems) are run by professional catering services.

Some school systems are going beyond the traditional one meal per day for school children to offer additional meals such as lunches for the elderly and hot breakfasts for children. The Hillsborough County (Florida) system permits students to help make planning decisions. It offers some à la carte choices on menus and is even adapting fast-food merchandising techniques. Still other schools have added student advisory boards to the food-service department.

Many public school systems face a series of problems that can cause difficulties for food-service sales people. One stems from the various buying practices that may be in effect even within the same system. For example, a centralized buying office may purchase many items through bids but allow individual lunchroom managers to purchase other items as they are needed. Other districts may require that most items, particularly food, be purchased locally.

Another problem concerns the difference in the calendar year, the school year, and the district's fiscal year. The confusion here, of course, is that the school year begins in one calendar year and ends in the next. For example, the school system may buy heavily at the end of its year (June) for delivery at the beginning of its next year (September) and have a third date for payments.

College and university food-service operations pose a different set of problems. On the college campus, these operations more closely resemble commercial feeding enterprises, particularly in food quality and variety. Many colleges have turned to commercial operators for student feeding, or at least for the management of college-operated facilities.

Purchasing for the varied services offered by colleges and universities originates from several sources. One is the school itself, either through its purchasing department or through a food-service manager or director. Another source includes food-buying groups, who may actually purchase for individual schools or simply advise and counsel the schools' buyers. State schools usually buy through the state purchasing department, with provisions for individual buying decisions by food directors. Commercial food-service operators who manage or run college feeding programs usually follow normal commercial procedures in their buying.

Airlines and Other Transportation

Pan American World Airways began its frozen food program in 1945 in order to be able to supply high quality meals, without regard to local availability of food ingredients or skilled labor. This was a novel concept at the time and in-flight systems have generally pioneered new food service programs. Since then the use of precooked or frozen ready-to-cook foods has become commonplace in the airline industry.

Because of space and weight limitations, the feeding operation must be carefully designed. Disposability is an additional factor to contend with. Airlines spend large sums of money to provide meals for their customers and are even able to offer various special meals for passengers. Both Kosher and vegetarian meals are available through caterers that supply the airlines.

The recent resurgence of railroad travel has brought with it renewed interest in feeding passengers. Taking a lesson from the airlines, the railroads have installed microwave ovens and are now providing their passengers with meals that do not require fully equipped dining cars.

The Institutional and Charitable Feeding Market

In both the commercial and the semicommercial feeding markets, customer satisfaction plays a major role in determining the characteristics of food-service operations. In the institutional and charitable (I&C) feeding market, customer satisfaction is usually subordinated to good nutrition or operational efficiency. For example, in hospitals and nursing homes, good nutrition is vital. Yet with soaring hospital costs, operating efficiency is almost as important as nutrition. The total emphasis on nutrition and cost in the I&C market is beginning to change, however, and I&C food-service people are also trying to add variety and interest to meals. This offers a challenge to dieticians and food-service managers as well as to food-service sales people. Sales people can help in planning menus, suggest different foods and packaging methods, and increase operational efficiency, but their advice must be based on a sound knowledge of the problems facing the I&C market.

The teaching hospital is a specific category of the I&C market because it combines educational and medical facilities, meaning that only about one-fourth of the food-service operation involves patients. Nearly three-fourths of the meals are consumed by the staff and visitors. This means that the food-service operation must prepare and serve not only the many special diets required by the patients but also more conventional foods. Indeed, the food-service operation of a large teaching hospital may easily approach the volume and the complexity of a commercial operation.

Quite a different problem is faced in another I&C category—the nursing home. A wide range of feeding facilities must be provided because nursing home residents range from the completely bedridden to the ambulatory. This range includes bedside service, individual table service in a central dining hall, vending service primarily for visitors and staff, and special-function services, such as social lunches and parties.

Another problem in a nursing home is the length of stay of residents, compared with the much shorter stay of hospital patients. To avoid monotony in meals, nursing-home menu cycles must be longer, say four weeks, compared to a two- or even one-week cycle in hospitals.

Of course, the range of food services supplied by nursing homes depends partly upon the type of care they provide. In those homes providing boarding care only, meals may be served family style, to help fulfill the social needs of the residents. In contrast, a home providing skilled nursing care is more likely to serve many meals (particularly breakfast) at the bedside but have limited dining hall service for ambulatory patients.

Government Feeding

Government markets are served only by suppliers who bid successfully on government contracts. Governmental markets include city, state and federal institutional and agency procurement authorities. Large by any measure, they represent great marketing opportunities and problems for firms. It is not governments that are sold to; it is their functional areas. In a major city, such as New York, an originator of buying situations includes education, police and recreation agencies.

It is not unusual to go into a firm today and find separate sales managers for industrial, consumer and government sales. Governments generally prefer to deal with manufacturers, although distributors do win some contract business.

12 CURRENT TRENDS IN PACKAGING

We are going to become more somber in our approach to our buying needs, and this in turn will generate a more sober approach to packaging. Many of the "frills" of the past will disappear.

William Gunn, of Stuart, Gunn and Furtha

The packaging industry entered the 1980s concerned about two new and vital factors in its potential growth: energy and waste. How to convince the consuming public that packaging does not contribute to waste was one of the industry's most pressing problems in the 1970s. As consumer advocates Barry Commoner and John Kouwenhoven have stated, the problems with energy, resources, and ecological insults have occurred not because of the ineptitude of technologists in reaching their goals, but because they have been so successful. The development of the plastics industry has created problems involving pollution and waste disposal. Huge paper mills have polluted waterways. Success in science has brought with it failure in the environment.

More government regulation in the 1970s led to more changes in packaging than in the previous thirty years, and this trend will only accelerate in the 1980s. Recycling, litter and bio-degrada-bility, terms associated with the ecology movement, have crept into the lexicon of the modern packaging technologist.

ECONOMIC TRENDS

Factors that will influence packaging trends in the 1980s include material shortages, inflation, increasing governmental restrictions, and changes in lifestyle. Economic factors will be particularly important in the future (see Table 12.1). While there has been a rapid growth in technology since the 1950s, the 1980s will probably see a decline in the use of materials already on hand, rather than an increase in new ones. Many new package concepts were introduced in the 1970s (see Table 12.2). It was an era of prosperity and one characterized by creative research. Consumerism became an important factor and many packages were designed to meet consumer demands. Plastic trays with visible contents, child-resistant closures, and shatterproof plastic beverage bottles are but a few of the packages developed in the 1970s to meet requirements of the consumer.

The packaging industry, under attack by both consumerists and environmentalists, fell back to regroup. Significant new forms included, in addition to the consumer-oriented packages mentioned above, the tin-free steel can for beer and the retort pouch. And new laminations of Surlyn and Nylon with older

TABLE 12.1   FUTURE ECONOMIC TRENDS AFFECTING PACKAGING

Trend	Effect on Packaging
Shortages	Existing materials will have to be used more creatively Less waste and overpackaging Increased use of cellulosics Fewer plastics
Inflation	Reduced developmental finances Greater use of natural materials Lag in capital investment and new-plant development Design to compensate for material expenses
Increasing governmental regulations	Greater scrutiny of new package concepts Increased government-supplier liaison

TABLE 12.2   PACKAGE DEVELOPMENTS: 1970–1979

Date	Type
1970	Child-resistant closures for aspirin bottles Larger bottles for soft drinks
1971	Foamed plastic jackets for glass bottles All-steel easy-open cans
1972	Waffle window trays for meat
1973	Easy-open cans with safety features Child-resistant aerosol overcaps
1974	Child-resistant foil pouches
1975	Retort pouches for institutional potatoes
1976	Metallized plastic-film coffee pouches (replacement for foil pouches) Ovenproof paperboard cartons Polyester soft-drink bottles
1977	Retort pouches are introduced
1978	Plastic cans for shortening
1979	Plastic carton for celery Composite cans for shortening Perforated ovenproof polyester pouches for french fries

films created a sort of semirigid material that could be sealed, shrunk, formed and cut to preference. But consumer fears that such packages produced problems with disposal, such as air pollution and litter, retarded full use of these innovations. Even Monsanto's Lopac bottle for Coca-Cola, which could be burned without polluting the air, was opposed by environmentalists.

Legislative control of packaging will continue through the 1980s, perhaps at a slower rate because of the Reagan Administration's deregulation strategy. Because of the combination of shrinking resources and competition from emerging Third World nations, shortages of many packaging materials appear to be imminent. There is simply not enough paper fiber available in the developed countries to supply the entire world. While the vast pulp potential available in Southeast Asia, South America and Siberia remains untapped, shortages of pulp appear inevitable in the very near future.

There are also finite supplies of petroleum, steel and coal. Glass, whose basic ingredient is unlimited, is in trouble because many Western states are opposed to the strip mining techniques used to obtain soda ash. Also, the severe natural gas shortages in 1979 indicated that the allocation of natural supplies may not be too far away. In packaging, questions such as which materials and how much of them should be exported may be foremost in the minds of legislators in future years.

The technical and regulatory aspects of packaging do not produce "consumer appeal." It is the happy marriage of the aesthetic and technical factors that often leads to the creation of a marketable package. While the engineer is fully aware of how to develop a package, he may not realize the design element inherent in making the package "sell." And, although the package designer knows graphics, he is seldom aware of the complex structural aspects involved in package fabrication.

DESIGN TRENDS

Changing trends in glass design was the subject of *New Glass—A Worldwide Survey,* an exhibition held at the Corning Museum of Glass, Corning, New York, in 1979. Several interesting glass bottles and vases that could lend themselves to consumer packaging were exhibited, particularly those that could be used as perfume bottles.

In 1977 Jovan, a Chicago cosmetics firm, introduced its "Man" and "Woman" cologne bottles. Marketed as a matching set, the bottles clearly symbolize sexual mating in both their shape and design. The designer, Pierre Dinand, created two bottles, each with a flat plane on one side and a curve on the other. This set is probably one of the boldest and most suggestive designs to appear on the packaging market. Product sales in the first three months totaled $15 million.

Another vibrant package is the "Macho" men's cologne bottle which was designed to have super-masculine appeal. It was considered imperative that the packaging say "Macho." The name was chosen because it suggests virility, assertiveness, and aggressiveness. Its T-shaped container is futuristic and phallic to some people and primitive to others.

Still another men's cologne package that is phallic is Pierre Cardin's. The bottle has a chrome-colored plastic cap and the package is a smoky-brown transparent plastic. Designed by Schwartz and Wassyng, it is probably more sexually symbolic than the "Macho" bottle.

Fig. 12.1 *A blown bottle, acid-etched, with applied decoration, by Hans T. Baumann, Germany, 1977. Courtesy, The Corning Museum of Glass, Corning, New York.*

Changes in interior design, furniture design, architecture, fashions, and personal lifestyles often influence package design. The growth of the disco culture has had its impact on style developments in several areas. Neon signs, metallic glitter and zany designs are appearing on the market.

Country-western music and the trend toward Western-style clothes is an even more recent trend. In some areas it has replaced disco as the dominant force. Perhaps that is why Ralph Lauren's "Chaps" cologne line is so popular. "Chaps" makes its appeal through "Western macho," reflecting the rancher in dungarees and leather jacket and chaps. This is in direct contrast to many other men's fragrances, which appeal to the image of the sophisticate.

There is now a growing tendency to intensify the impact of the package through color and bold shapes, either sharply geometric forms or sculptural shapes. Several packages on the market offer a view into the late 1980s. Among these are "My" perfume by Myrurgia and Geoffrey Beene's "Red" fragrance line.

The "My" line by Myrurgia was created by the French sculptor and designer Pierre Dinand, using a powerful red-black-silver color combination. The bold interplay between squares and triangles is used on the outer carton and in the design of the bottles and caps. Geoffrey Beene's "Red," consists of various-size perfumes packed in a brilliant red lacquer box, uses one color to make a strong statement. The red bottles are slender, elongated, vertically faceted, and topped with a black-tasseled cap. One item in the line, a crystal sphere, has a flat, square stopper that counterpoints the shape of the bottle. Other package designs that explore the interrelationships of shapes, particularly squares and circles, include "Azzaro" by Prince Matchabelli, "Symboise" by Stendhal, and "Lamborghini" by Ted Lapidus.

Fig. 12.2   Packed Cans, *by Christo, 1958; cans, lacquered canvas, and twine. Courtesy, the artist; photograph by Eeva-Inkeri, New York City.*

Fig. 12.3 *Norman Rockwell's oil painting*, Ballantine's Ale, *18 x 22 inches.* Courtesy, *the collection of John B. Walborn and reprinted with permission; photograph by William Doyle Galleries, New York City.*

The Retort Pouch

Hailed by many as the most important development in food packaging since Nicholas Appert's discovery of food preservation, the retort pouch has been advertised by ITT as "No utensils, no cleanup. It's even easier than eating out."

The idea of marketing a shelf-stable, flexible pouch for food has been debated since the late 1950s. In 1960 the Army produced shelf-stable retort pouches. For a soldier, the retort pouch is a natural. The package weighs nothing and is easily disposed of in war zones. During World War II, cans were colored olive-green so they would not reflect light rays when they were discarded. In a retort pouch this is not necessary. But lack of suitable technology precluded widespread distribution of the pouch. Now the time appears ripe for wider use of the retort pouch both for food and drugs. New plastics are available, processing technology is more sophisticated, and most of the FDA hurdles have been overcome. It now seems that the institutional-size retort pouch will lead the way for the smaller size.

Previously, FDA and USDA approvals were slow to come, and sometimes they were withdrawn. Initial approval for its use in the United States came from the USDA in November 1974. Only a few months after USDA approval was issued, the FDA ruled that the retort pouch was not in compliance with two pertinent FDA regulations although industry had long been convinced that it did comply with FDA regulations.

As a result of the FDA action, the USDA approval was cancelled and, at the direction of the FDA extracts of the adhesive used by all retort material suppliers for bonding the sealing ply to the foil were required to be fed to two animal species to demonstrate their innocuousness. Those feeding tests have been completed only recently, and the results are currently being evaluated.

The pros and cons for the retort pouch are summarized in Table 12.3.

The retort pouch is currently used to package a variety of foods in Japan and Europe. Japanese sales amount to more than 400 million units annually, while European processors sell about 45 million units annually. The Japanese success story is a consequence of their rising leisure class. The rise of an affluent population in a country with little or no refrigeration became a prime target for staples packaged in retort pouches. In the European market, sausages were the basic commodity packed, but the retort pouch can also supplement the Europeans' choice of food

TABLE 12.3   THE RETORT POUCH

Advantages	Disadvantages
Container costs less	Major capital investments necessary
Rapid heat penetration, so less energy is needed for processing	Processing requires special racks and superimposed pressure
No heavy metal problems (particles on tinny-tasting foods)	Limited in size; if too large, profile gets thicker, requiring more processing
Easy opening	Overwrapping necessary for product protection
No disposal problems	More fragile and subject to puncture and breakage

products. Normeats "Royal Dane" products (1967), Howard Redi Foods (1969), Star Foods (1969), and others now market retort products in Europe.

In North America both ITT Continental Baking and Swan Valley Foods, a Canadian packer, have introduced a line of retort products. An even more recent addition has been the Kraft Foods line of entreés. FDA-approved materials are now available from the Reynolds Metals Company, the Continental Can Company, and the American Can Company.

There are several basic pouch requirements:

1. Heat seals around the pouch must be sound
2. Heat seals must endure high temperatures
3. Must withstand abuse
4. Must not be readily puncturable
5. Must protect against bacteria
6. Must protect against moisture loss

Considering the availability of materials and the limitations imposed by manufacturing equipment in the flexible packaging industry, the three-ply type of structure bonded with polyurethane adhesives appears to be the optimum lamination for the retort pouch so far as price is concerned. Other laminations include polyester/foil/polypropylene and polyester/foil/high-density polyethylene. However, these are not the only materials that could be used. Because of the FDA setback, many companies, particularly in the United States, set out to develop other flexible laminations that would comply with existing FDA regulations and avoid the necessity for animal-feeding studies. Continental Can Company was the first to do this, but Reynolds Metals and American Can introduced a fully approved material soon afterward. The new retort pouch for food still has a polyester-face ply, an aluminum-foil center ply, and a polyolefin sealing ply, but it differs from the traditional pouches in that the polyolefin sealing ply is bonded to the foil.

Despite the twenty-year lag between the development of the retort pouch and the commercialization of a new version for foods in the United States, millions of the former pouches have been used for packaging thermally sterilized hospital supplies, medical devices, and pharmaceuticals.

The retort pouch will become a dynamic package system for underdeveloped countries and areas of high population density, and it will be seen by consumers as an alternative to frozen foods because less energy is required to store the product. Manufacturing speed will increase and costs will decrease. Still required is increased consumer education and industry's willingness to fully evaluate the concept's potential.

Japanese Trends

The close association that the Japanese have with nature historically has been reflected in their packaging. Bamboo, straw, clay, and wood have often been used as packaging materials for Japanese foods and drugs. In his book, *How to Wrap Five More Eggs* (Weatherhill, 1978), Hideyuki Oka, a leading graphic designer in Japan, laments the disappearance of traditional packaging in Japan. In 1975 he stated, "Japan today is a far cry from the world symbolized by its traditional packages. That world simply no longer exists except perhaps in a few inaccessible and forgotten spots." Traditional Japanese packages were the outgrowth of practicality, beauty, folk handicrafts, and an "aesthetic consciousness of propriety," reflecting a love of spiritual things.

Fig. 12.4 *An Odette candy box; a design treatment of a Tamatebako (treasured casket image) produces a dreamy, romantic feeling. Courtesy, Morozoff Chocolate Co., Ltd., Kobe, Japan.*

Glass and other packaging materials were exhibited in the 1980 show in New York entitled "Packaging Graphics in Japan." Sponsored by the Japan Package Design Association, the exhibit offered a rare glimpse into contemporary oriental packaging. The exhibit featured traditional packages constructed from natural products, but the overwhelming majority of the packages were more Western in design. Particularly interesting were two biscuit tins—"Arcadia" and "Odette." The Odette tin evokes a brilliant Art Nouveau image. Another example, of the new Shogun sake, is magnificent in form and figure. Other designs of future interest include the Shiseido-Zen perfume bottle, the Suntory Royal whiskey bottle, and Cook Do seasoning mixes in retort pouches.

APPENDIX THE HISTORY OF PACKAGING

Speculation about the first package makes one wonder whether Eve's second apple was wrapped in grape leaves tied with a vine.
Janet S. Byrne, *Metropolitan Museum of Art Bulletin (1976)*

THE PACKAGES OF ANTIQUITY

Before pottery was invented, Stone Age man had to improvise his containers from whatever nature provided. And his possibilities were limited: large scallop shells found near the sea, or in the tropics, wild coconuts, banana leaves, gourds, hollow tubes of bamboo. (Indeed, the availability of bamboo, may have contributed to the development of distillation long before the birth of Christ.) But wherever man was to be found, skulls were also found, and man discovered an excellent, almost indestructible cup. Broken skulls such as those found in the caves of Nyandong in Java could well be considered to be the first packages ever used by man. Centuries later, the nomad Scythians still made cups from the skulls of their enemies. Herodotus IV wrote: "They saw off the part below the eyebrows and after cleaning out what remains stretch a piece of rawhide round it on the outside. . . . When important visitors arrive, these skulls are passed round and the host tells the story of them."

Pottery

The invention of pottery was one of the most important achievements of the Neolithic period (6200–1400 B.C.) Now man could cook and store his foods in fireproof, watertight containers. Early pottery was built in successive rings of coils, which were often scraped smooth and burnished with a pebble. The earliest known painted decorations on pottery were found in Cilicia, in southwestern Turkey, about 1850; pottery was covered in cream slip and motifs painted in red clay. Pottery soon found scores of new uses, from carrying water to holding wines and perfumes. The technique of decoration progressed from primitive Neolithic paintings to the high-gloss Greek pottery.

Perhaps the widest application of pottery as a packaging material came about because of its usefulness in shipping and trading. At the time, the world depended on trade to maintain its food supply. Led by the Phoenicians, the Egyptians, Greeks, and Romans all traded wheat, wine, olive oil, spices, and perfumes in pottery containers. And what better way to identify the contents than to decorate the jugs, urns, and vases.

Urns were often buried in tombs as offerings to the dead. The first of these were discovered in the 18th century in the Etruscan tombs in Italy by archaeologists. The Greek vases found in Italy

Fig. A.1 *Unusual Egyptian Painted Wood Rectangular Mummy Case for a Serpent, 2½ by 7½ inches, possibly Late Period. One side has a winged heart scarab, the other a winged zoomorphic figure flanked by two standing figures. The ends are decorated with various figures; the sliding top is surmounted by a bronze serpent of a later date. Inside is a mummified serpent. Courtesy, Christie's East, New York City.*

had been sold in ancient times to the Etruscans as export merchandise by Greek traders.

The importance of the pottery container in trade was recorded by Herodotus in 530 B.C. when he wrote that the Persians supplied conquered Egyptian cities with water and wine in earthenware jars. These were collected later and returned for reuse. In Greece, as in Egypt and Palestine, unguents were kept in jars made of alabaster and onyx. And in China there were Neolithic mortuary urns and finely decorated wine and grain jars from the Han period (206 B.C.–A.D. 220). Although the ancient Greeks decorated their vases with people, the Chinese preferred animated scenes and fantastic mythological monsters. In Mexico, the Oaxaca urns made by the Mixtecs and Zapotecs featured highly ornate figures.

Up to the mid-17th century, nearly all bottles for wine or oil were made either of leather or earthenware; the only exceptions were a few small bottles and vials used for medicine.

Glass

In the 1st century B.C., glass blowing was discovered. Glass is one of man's most versatile creations. Stable and durable, it is obtained by the fusion of silica (sand/quartz) with an alkali (soda)

Fig. A.2 *The Northampton Vase, c. 540 B.C. A clear red-brown clay vessel embellished with black Triton figures holding aloft a white dolphin, Dionysos, satyrs, and pygmies riding cranes. Acquired by the second Marquess of Northampton in Italy in the 1820s, it sold at auction at Christie's in London in 1980 for $448,400. Property of a private collector; photograph courtesy of Christie's, London.*

and calcium carbonate (lime). It can be engraved, painted, enamalled, gilded, ground, cut, and shaped, and colored by adding metallic oxides. The effects are infinite, and the result is a blend of both the practical and the aesthetic.

When and where this art began remains controversial. There is now much evidence that glass originated in Asia Minor and its early manufacture concentrated in Egypt and Mesopotamia. The first clear picture of a glass industry emerges only with the 19th Dynasty in Egypt (1570–1304 B.C.). Initially, small vessels were produced by enclosing a sandy core in a layer of usually opaque-blue glass and decorating it by applying threads in constrasting colors. This was the main process used in the Mediterranean until the 1st century B.C. Egyptian glass containers were used to hold cosmetics, ointments, and perfumes.

In A.D. 14, a glass-makers guild was established in Rome and the specialist artisan was fully recognized, although it was not until the 3rd century that glassware was mass-produced. Roman glass is noteworthy because of the wide variety of techniques that were developed for forming and decorating it. It was often blown into molds and decorated with wheel-cut lines, pinched-out projections, applied blobs, and dimples. The natural color of glass is pale green, the shade varying with the raw materials used, but other colors can be produced by adding different minerals, and it

Fig. A.3 *Roman glass, 2d century. Left: a cylindrical bottle in pale green with a slightly indented base and heavy disc rim, 5¾ inches (146mm). Right: a pear-shaped bottle, in pale viridian, 5 inches (127mm), has a slightly indented base; the rim is folded and flattened. Courtesy, Charles Ede Ltd., London, England.*

Fig. A.4 *Roman Phials. Left: Royal blue phial, 4½ inches (114mm), c. 3d century, from Mount Carmel. The central swelling was lightly fluted by using a tool rather than a mold. The lip is folded and slightly flattened. Right: Yellow glass phial with a cylindrical body tapering to a peg base, 4⅜ inches (111mm), 1st century. The long neck has a flared lip. Courtesy, Charles Ede Ltd., London, England.*

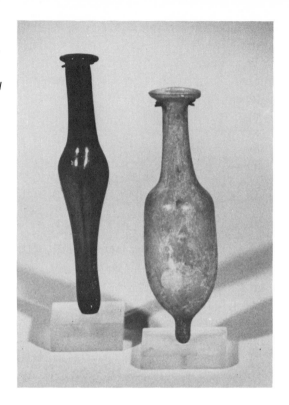

can be rendered colorless by adding antimony or manganese. A vogue for colorless glass seems to have begun toward the end of the 1st century. A century later, most of the finest glass was colorless. Roman bottles of this period often had embossed decorations, sometimes with the maker's name or an advertisement. Glass bottles in England were still a novelty as late as 1650.

Leather, Wood, and Metal

For cooking, primitive man used the stomachs of the animals he killed for food. Here was a container that was waterproof and heatproof enough to be hung over a fire. As late as the 5th century B.C., the Scythians cooked their food in a stomach bag. By about 1300 B.C., leather-working techniques had improved, so skins began to replace many of the older containers. The leather bag as a wineskin was referred to by Jesus of Nazareth (*Matthew*

9:17): "Neither do men put new wine into old wineskins; else the skins burst."

The barrel was invented by Alpine tribes, according to Pliny. Its carefully fitted staves and heads were fashioned from wood, interlocked by mortises and the whole bound together with iron hoops. Wooden chests were made in similiar fashion. And while boxes and cups were made from silver and gold in ancient times, they were too valuable for common use.

The Romans also used lead in many ways, including water pipes, but the only known evidence in packaging is their application of lead seals to ointment jars. It is likely they found ways to hammer the soft metal into thin foils. They may have wrapped foods in it, not knowing it was poisonous.

Precious metals had long been used in box making, but about A.D. 1200, artisans in Bohemia discovered a hot-dip process for plating tin on thin sheets of iron that had been hammered out by hand from rods. Although tinplate had been discovered, there are no records of tinplate containers until the late 1700s, when tobacconists in London began selling snuff in metal containers.

Paper

The Egyptians discovered papyrus about 3000 B.C., but paper as we know it was first made at Lei-Yang, China, by Ts'ai Lun in A.D. 105.

Paper making had originally come to Europe with the Arabs, who had picked them up when they overran Samarkand in A.D. 751, just after the Chinese had sent a team of paper-makers there to set up a factory. By 1050 the Byzantine Empire was importing Arab paper, and in Europe paper was first made in Muslim Spain at Xativa, north of Valencia. The first water-powered paper mill was located in Fabriano, Italy, in 1280.

MEDIEVAL PACKAGING

The art of bag making out of paper dates back to at least the Thirty Years' War (1618–48). Although the earliest surviving printed paper package, attributed to Andreas Bernhart, a German papermaker, is from the 1500s. Before the paper bag, it was not uncommon for royalty to promenade with resplendent *aumônières,* the purse. Carried with pomp and ceremony by manservants, these bags were symbols of wealth and station. Chaucer describes the "aumônière" of the carpenter's wife as "tasseled with silk and pearled with latoun" (brazen knobs). Even after the advent of the pocket in the sixteenth century, the socially prominent wore embellished "aumônières" on their belts. The Middle Ages also produced the *gipciere.* Worn at the side, it was a bag with three tassels and a bulging front. Made of leather, silk, or velvet, it had an ornate frame inlaid with silver on which were inscribed moral and pious phrases.

There is little information as to the quantities in which goods were bought in medieval England, but it is certain that nothing

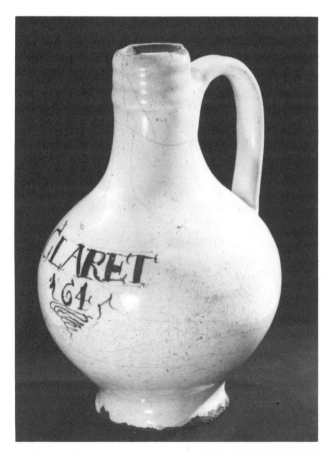

Fig. A.5 *Lambeth Delft tin-glazed earthenware wine jug, c. 1645. This was in common use before the reintroduction of glass in England in the 17th century. Courtesy, Harvey's Wine Museum, Bristol, England.*

was wrapped or packaged. The purchaser provided his own containers, such as baskets, jugs, bowls, and sacks. It is believed that the butterbur got its name because its great, thick leaves were once used to wrap small quantities of butter. In the 17th century, widespread use of glass bottles for wine and the importation of tobacco and tea gave birth to containers that are now valuable collectibles. These and scent bottles, used to disguise the odor of not bathing regularly, were the only packages used up to the end of the 18th century.

At the dawn of the Industrial Revolution most individual packages used were for the expensive commodities of the day—wine, tobacco, tea, medicines, perfumes, and even gunpowder. Almost nothing sold in England was branded, packed, standardized, or priced and customers had to be on the watch for shopping was a

Fig. A.6 *Wine bottle, c. 1720–30. By 1720, distinct straight sides had replaced curved ones, and the bottles were wider at the base than at the shoulder. Courtesy, Harvey's Wine Museum, Bristol, England.*

risky business. Individual packages of medicines began to appear. Medicinals were sometimes sold in "sealed" paper wrappers in the 1700s, and one could even buy papers containing pins, tobacco, tea, and wig powder.

THE MACHINE AGE AND THE INDUSTRIAL REVOLUTION

The Machine Age and the Industrial Revolution set the course for the future development of packaging. Before the Machine Age, which began in the latter half of the 17th century, most articles of manufacture were the product of hand labor. The word itself comes from the Latin "manu" (by hand) and "factus" (made). Only simple machines powered by muscle, gravity, or water were available. The foot treadle and crankshaft turned the spinning wheel, the potter's wheel, the grindstone, and the carpenter's lathe. Dogs were trained to run on treadmills to turn roasts on spits. Horses or oxen were harnessed to windlasses to move great loads. Windmills pumped water, and waterwheels turned gristmills or powered sawmills.

Newcomb and Watt brought the steam engine to practical reality in the 1700s, while Lavoisier, Dalton, Priestley, and others were building the foundations of modern chemistry, and Galvani, Volta, and Ohm were investigating the mysteries of electricity.

Mass production, standardization of parts, and power-driven machinery were combined successfully in 1700 by Polhem in Sweden; similar manufacture was carried on in France in 1762. In 1785 a French gunsmith named LeBlanc showed the process to the American ambassador, Thomas Jefferson. Eli Whitney pioneered the procedure in the United States. In 1803, after two years of tooling up, Whitney's gun factory in Hamden, Connecticut, began producing muskets at three times the rate skilled gunsmiths could make them by hand. Before long, mass production came to be known as the "American system."

The search for more efficient methods of production led to the invention of more and more labor-saving machines and even to machines for *making* machines. The new machines created a need for new materials of construction. Wooden parts couldn't stand great strain or long wear. Brass was expensive. Cast iron was hard but brittle. Wrought iron was tough but too soft. The only available steels, made from wrought iron, were expensive. In 1856 Henry Bessemer invented the process of converting pig iron into steel. This not only made the metal plentiful and cheap, but it created medium carbon steels, which were perfect for long-wearing machine parts.

The textile industry was one of the first to convert from hand labor. In rapid succession from 1733 came Kay's flying shuttle, Hargreave's spinning mule, and Cartwright's power loom. These machines could spin and weave stronger thread much faster than hand labor. They could process pure cotton fiber instead of mixtures, and thus cotton became king.

During the 1800s new machinery, new processes, and new scientific discoveries occurred at a staggering rate:

Prime Power: Steam engines, internal combustion engines, the dynamo, the electric motor

Communications and Graphic Arts: Telegraphy, photography, lithography, retogravure printing, the telephone, the typewriter, the electric light

Transportation: Steamships, steam railroads, the automobile, the airplane, the bicycle, the hot-air balloon

Medicine: Bacteriology, antiseptic surgery, anesthesia, chemical therapy, immunization, sanitation

Food and Agriculture: Food processing, seed planters, cultivators, harvesters

Clothing: Sewing machines, power spinning, power weaving

Primary Industry: Steel, petroleum, coal, coke, coal-tar products, soap, rubber, pulp and paper, plastics

The Industrial Revolution and the resulting immigration to the cities imposed severe demands on society. On the farm there was adequate storage for a good supply of food that needed to be preserved, cured, or dried. In the towns, however, storage problems meant the consumers had to purchase products in smaller amounts, and this led to smaller, individual packages. In England, patent medicines began to appear in the late 18th century,

Fig. A.7   *This can of roast veal was taken on Perry's voyage in 1824. It was opened 111 years later, and its contents fed to rats with no adverse effects. Courtesy, International Tin Research Institute, Middlesex, England.*

and many of these early packages were exported to America. These English consumer packages, which are collectibles today, include patent-medicine bottles, labels, and stoppers, as well as the novel biscuit tins of mid-century. The first wrapping paper in rolls was produced in London in 1807 and in Delaware in 1815. In the United States, early Shaker packaging began to appear in the late 1790s. Because the Fourdrinier's machine, developed in France, could produce cheap paper, packaging was on the threshold of a new era. In 1827, the first Fourdrinier machine in the United States was installed in Columbia County, New York. This machine made it possible to produce a variety of paper products on a large scale, including paperboard, labels, and paper bags. There is some dispute as to when the first bags were made. The making of bags out of paper dates back to the years 1618–48. The first English paper bag manufacturer was established in 1844 in Bristol, and Francis Wolle developed the first bagmaking machine in 1852 in the United States. The first machine capable of producing square-bottomed bags appeared around 1870. Bag production took a giant leap forward when the paper bag replaced cotton bags during the Civil War.

Another significant milestone in the history of packaging came about in the early 1800s with the introduction of the tinplate can. Then it was possible to make 60 cans a day; today, 1,200 can be made per minute.

The development of lithography by Alois Senefelder in Munich, Germany in 1798, corrugated board in England in 1856, offset lithography in 1875 by Benjamin George in London, and flexography by C. A. Howley in France in 1890, gave rise to printing techniques. These new printing techniques, coupled with the introduction of new packaging materials and concepts, gave the 19th-century package a distinctive flair.

But despite all these new developments, there was still the problem of getting the product to the retailer in usable condition.

PACKAGING IN VICTORIAN ENGLAND

Queen Victoria's reign, which began in 1837, saw the mechanization of the infant packaging industry—the industrialization of tin cans, wooden barrels, paper bags, and glass bottles—and gave rise to package designers, industrial designers, and new package concepts. England was rapidly becoming the workshop of the world. Years later, in 1916, Elizabeth Hardwick wrote in the Victorian Society *Bulletin* that "The raging productivity of the Victorians . . . was a thing noble, glorious, and awesome in itself."

One of the greatest boosts to trade and packaging was the Crystal Palace exhibition of 1851, which marked the culmination of the changes initiated by the Industrial Revolution. The application of technology in industry produced enormous changes, not only in material comforts but in the very foundations of civilized life.

The Crystal Palace, the brain child of Prince Albert and architect Joseph Paxton, resembled a huge shop window with products from many countries and more than 6 million people flocked to Hyde Park to see it.

Out of the Exhibition came Whiteley's idea of a "department" store. (A. T. Stewart developed the same concept independently in New York in the 1870s). William Whiteley started his first store in Bayswater, London, in 1863 with a staff of three. Later to be called the "Universal Provider," he opened a jewelry department in 1867, and thus became the first British retailer to break away from drapery and haberdashery in a departmentalized store. A "foreign department" selling Oriental novelties followed three years later, then an estate agency and a restaurant (the first in a store) in 1872. Whiteley was eventually murdered in 1907 by a man who claimed to be his illegitimate son.

The Great Exhibition also marked the beginning of active promotion of glass packaging. The leading exhibitor of bottles was E. Brefitt of Castleford; more than one million bottles of "pop" were drunk by the thirsty crowds. There were also some printed mental boxes exhibited. These were decorated by transfer printing; the decoration was transferred to sheets of paper that were pasted on the box, then varnished.

Fig. A.8 *The decorated ceramic pot lid began to emerge in England about 1850. Courtesy, Landor Associates, San Francisco: Museum of Packaging Antiquities.*

New developments in glass-making soon followed. In 1872 two important British developments in closures were Barrett's internal screw stopper and the Codd bottle, sealed with a captive glass marble inside the neck. The Codd bottle is still used today in the Far East.

In 1880, William Ashley in Castleford invented the world's first semiautomatic bottle-making machine. And when Michael Owens invented the automatic bottle-blowing machine in the United States in 1903, the entire complexion of the glass industry changed. By 1910 British glass-bottle production had been mechanized on a large scale. Tin cans were still made slowly, and semiautomatic can-making machines were not installed until 1927.

The first plastic, "Parkesine," was developed about 1856. Made by Alexander Parkes of Birmingham and originally manufactured by the Parkesine Company at Hackney Wick, London, it was a thermoplastic material produced from nitrocellulose, camphor, and alcohol. Exhibited at the Great Exhibition in South Kensington of 1862, it was described by a contemporary observer as "the product of a mixture of chloroform and caster oil which products a substance hard as horn, but as flexible as leather, capable of being cast or stamped, painted, dyed or carved and which above all, can be produced in any quantity at a lower price than gutta percha" (Prize medal leaflet, International Exhibition, 1862). The juries of the exhibition awarded the inventor a bronze medal for excellence. Little did they know that they were saluting the birth of a great new industry.

In the late 19th century, developments in plastics accelerated. M. Berthelot synthesized styrene in 1866, John Wesley Hyatt's basic celluloid patent appeared in 1870, and in 1872 Hyatt patented the first plastics injection molding machine. In 1872 Hans Baumann reported the polymerization of vinyl chloride. And in 1894 Cross, Beavan and Beadle produced the first industrial process for the manufacture of cellulose acetate.

In 1836 the government-sponsored Committee on Arts and Manufactures established schools of design throughout England, and in the 1840s and 1850s, a new cadre of package designers began to appear on the scene.

Many early package designers borrowed their ideas and inspirations from a pattern book *The Grammar of Ornament* (reissued by Van Nostrand Reinhold, London and New York, 1972), published in 1856 by Owen Jones, an interior decorator who chose the color scheme for the interior of the Crystal Palace. *The Grammar of Ornament* contained 3,000 designs from China, the Far East, the Middle East, and India, as well as a range of Western patterns. The Murad cigarette package of 1900 is a dramatic example of the influence of Jones's book on package design. The artist borrowed heavily from the Egyptian plates found in "Grammar of Ornament". Owen Jones also designed the Huntley & Palmers "Casket" biscuit tin of 1868, commemorating the granting of royal patronage to the company. This tin is probably the earliest surviving example of tin printing to which a definite date can be ascribed in England.

For British commercial artists, even more examples of ornament were available, such as Pugin's *Glossary of Ecclesiastical Ornament* (1844) and Wyatt's *The Art of Illuminating* (1860). The reputation of the British chromolithographic industry was so strong in the mid-1800s that even the celebrated French poster artist Jules Cherét came to London to learn more.

The Victorian package design industry filled shelves throughout England. Prepackaged commodities, such as tea, were now readily available to everyone, and no longer was the shopper dependent on limited local supplies. *Mrs. Beeton's Book of Household Management,* published in 1861, gives advice about marketing, and the author appears to have taken it for granted that her readers would be able to buy everything they needed. She refers frequently to tins of this and packets of that in her lists of recipe ingredients.

THE GILDED AGE IN THE UNITED STATES

The importance of packaging in this period is summed up by Robert Atwan, Donald McQuade, and John W. Wright in *Edsels, Luckies and Frigidaires* (1979):

> The mass produced food package, surely one of the most significant innovations to come out of the Gilded Age, depended on two related phenomena for its sales effectiveness:

Fig. A.9   *Early Colman's mustard packages.* Courtesy, *Colman Foods, Norwich, England.*

(1.) the enactment of tough trademark protection laws, and
(2.) a power shift throughout the entire commercial structure that enabled the manufacturer, rather than the retailer, jobber or wholesaler, to determine the where, what and when of mass consumption.

When Queen Victoria ascended the throne in 1837, Andrew Jackson had just concluded his term of office and Martin van Buren was President. The United States had been gradually reducing its imports and dependency on England. Indeed, after the War of 1812, manufactured product imports swiftly declined, and by 1850 the major wholesale centers in the United States were New York and Chicago.

Although as early as 1844, Colonel Dennison had started the hand manufacture of rigid set-up boxes on a commercial scale, American packaging in the 1850s was not as advanced as English packaging. Veneer baskets were commonly used for produce, glass or earthenware vessels were used to ship pickled oysters and olive oil to the West Indies, and gunny bags held grain. But perhaps the most important container of the day was the barrel. Flour, rum, molasses, whale oil and other liquids were transported in barrels. The 1850 census records more than 43,000 coopers, and by 1860 there were 2,700 cooperages and 300 companies that made staves, headings, and hoops. As early as 1811, a machine to shape barrel staves was patented, and dozens of alternatives to the 1811 design were patented over the years. At the New York Crystal Palace Exhibition in 1853, three stave-making machines were among the silver-medal winners, as was a display by an Elmira, New York, inventor of keg and barrel machinery.

The first collapsible tube was patented by artist John Rand in 1841 and was first used commercially by the DeVoe and Reynolds Company for packing oil paints. The first American tube manufacturing plant started production in 1870. In England, Rand's lead tubes were soon adopted by London artists. This jump in American ingenuity probably led to the development of the first toothpaste tube in 1892 by Dr. Washington Sheffield, a New London, Connecticut, dentist. These tubes were then later manufactured by his Sheffield Tube Corp. Toothpaste had previously been packed in round pots, usually of Staffordshire ware.

Although the expansion of the railroads opened up new markets for manufactured products, it was the Civil War that catalyzed the growth of mass-production technology. Paper tubes were first used to package ammunition. Armies had to be fed and clothed, giving rise to fast-speed production and sophisticated manufacturing methods. The glued paper sack replaced the cotton flour sack. George West, a New York mill owner, developed a paper sack that could hold fifty pounds of lead.

The peddlers who provided American homes with various commodities began to disappear and did not reappear again, as railroads brought people nearer to markets and stores. In 1850, there were 10,669 peddlers on the road, and in 1860 there were 16,600. But President Grant's notorious "anti-semitic" order of the Civil War did much to decrease the peddler's activities.

In 1879 Robert Gair developed the first folding carton. Gair, a paper jobber and converter came to the United States from Edinburgh, Scotland in 1852.

When a printer set the type rule too high, creased the paper, and ruined a job, the printer thought he would be fired. But Gair saw the potential of the creased score. He eventually built the Gair Industrial Village in Brooklyn and later in Piermont, New York.

There were many other outstanding developments in the late 19th century. Chromolithographic printing produced brightly

Fig. A.10   *Pint glass milk bottle, c. 1900. Courtesy, Landor Associates, San Francisco; Museum of Packaging Antiquities; photo by Jeanne Riemen.*

colored paper labels and overwraps. A.L. Jones received the first U.S. patent for corrugated paperboard in 1871. Three years later, Oliver Long took another giant step forward with a process for sandwiching it between paperboard sheets. These events gave birth to a burgeoning new industry—corrugated containers. The first milk bottle was introduced in 1879 by Echo Farms Dairy Company in New York. Known as the "Whiteman Milk Jar," it was quickly succeeded by the first embossed milk bottle, "The Thatcher Bottle." Dr. Henry D. Thatcher, a Potsdam, New York, druggist invented his milk bottle with a closure similar to that used on his Lightening fruit jar. In 1886 he developed a glass closure for his bottle, and in 1889 he revolutionized the industry by introducing the "Common Sense Milk Jar," which had a groove inside the bottle upon which rested a waxed-paper cap. (Aluminum-foil bottle caps were first produced by Josef Ionsson of Linkeping, Sweden, in 1914.) Thatcher noted that the cap meant "no rusty metal covers or twisted wire fasteners, less breakage and [the bottle] can be washed absolutely clean and much quicker than any other . . . avoiding tainted or sour milk" (W.C. Ketchum, Jr., *American Bottles,* 1975).

Other 19th-century developments in packaging included the use of tinfoil in food packaging. In New York City, the John J. Crooke Company had been rolling tinfoil as early as 1855. Wax paper was first introduced in 1877, when a candlemaker, tired of carrying fresh fish home from the market in newspaper, founded a company to make waxed paper, coating newspaper with candle wax.

At the turn of the century, paraffin waxed paper was being used as an inner liner for cracker cartons and for candy wraps, then as a bread wrapper. In 1902 rolls of waxed paper in a cutter-edged carton were produced.

The 1890s also witnessed the early beginnings of the plastic-film industry. Soluble cellulose was first produced in the laboratory in 1892, and by 1898 a method for making thread and film from viscose had been introduced. These two steps led eventually to the development of cellophane.

Also in the late 1890s, one of the strangest incidents in the history of packaging occurred. Rags were still an important and expensive ingredient of high-quality paper. An American paper manufacturer named Augustus Stanwood purchased a large shipment of bandages stripped from stolen mummies to use in the manufacture of rag paper. But most of the wrappings were stained with resin, which could not be bleached out. To preserve his investment, Stanwood used the material to make ordinary brown butcher paper.

From 1850 to 1900 the American packaging industry witnessed the introduction of new machinery capable of mass-producing many packages that were previously made by hand. Barrels, baskets, tin cans, glass bottles, paper bags and fibreboard cartons began to be manufactured in commercial quantities.

Largely used to contain fresh produce, the veneer basket and berry box was first mass-produced in 1840. The machine used a long, sharp knife edge to peel a continuous sheet of veneer off of a rotating section of a log, which was mounted between centers and turned against the knife by water or steam power. The veneer was then cut into pieces for the body of the basket; a solid piece of board was used for the bottom. By 1860, the veneer lathe was being used, and by 1900 almost all produce baskets and other wooden tubs were machine made.

The tin can industry was mechanized in about 1880. The Cox hand capper gave the industry its biggest push into the future, and developments rapidly followed, such as Norton's continuous can-making line and the wide-mouth "sanitary" can of the early 1900s. In 1901 the American Can Company was established, merging 100 tin-can manufacturers.

Until about 1890, all bottles or food jars were made by skilled glass blowers using methods that had not changed fundamentally for hundreds of years. Glass-bottle mechanization began in 1881, when Philip Arbogast, a Pittsburgh glass maker, received a patent for a "press and blow" two-stage mold. In 1896 the semi-automatic machine for the regular production of wide-mouth jars was introduced. Bottle automation reached its peak in 1903, when Michael Owens invented the first fully automatic bottle-making machine in Toledo, Ohio.

In 1850 pulp became available in large quantities. Paper bags began to appear in the 1850s, and the cutting and scoring of small paperboard boxes could be done on job printing presses that were plentiful by the 1880s. In 1875 about 600 million bags were being produced annually. And in 1883, the "automatic" or self-opening bag was introduced.

MODERN PACKAGING EMERGES

When paperboard became popular because of mass production in the late 1860s, many new types of products were packaged, including hat pins, wearing apparel, and sewing equipment. The first use of paperboard for food packaging was by the National Biscuit Company, and paper bags soon replaced bulk cloth sacks used for flour, potatoes, and onions. Changes in technology and lifestyle in the early 20th century were reflected in even more changes in packaging. A national distribution system began to fall into place as railroads moved goods from coast to coast. Chain stores sprang up, and the package became an important marketing tool. No longer did consumers have to rely on a salesperson for help in selecting the product: They could select what they wanted directly from the shelves. Indeed, the supermarket itself is an outcome of the growth of the small, unit-packaged food products.

By 1920 the stage had been set for the radical changes that occurred in future decades. Supermarkets were beginning to appear, refrigeration became widespread, and more and more people had automobiles. The most striking development in packaging to come out of the early 20th century, was the use of transparent film. Flexible, transparent, and ultimately moistureproof and heat-sealable, it has had an important impact on packaging. Sophisticated variations in film packaging have made this material a continuing innovation in packaging design.

Fig. A.11 *Chewing tobacco pouch, c. 1900, an early example of tobacco packaging.* Courtesy, *Collection of Edward Morrill, Werbin & Morrill, Inc., New York City.*

Advancements in retailing and marketing led to significant developments in packaging technology. The packaging revolution, which began in the 1920s, was called "one of the most manifold and least noticed revolutions in the common experience" by Daniel J. Boorstin (*The Americans: The Democratic Experience,* Random House, 1973). As late as 1928, 90 percent of all sugar was still sold in bulk. Advertising people recognized the promotional value of attractive and distinctive containers. The Great Depression gave added impetus to efforts to supply the demands of consumers and, through advertising, to expand and intensify these demands. The notion that a distinctive package could help to sell a product preceded by some years the appearance of self-service supermarkets, but it was clearly in the supermarket that the package as a sales tool became firmly entrenched.

Frozen foods were introduced to the public, and the home-freezer began to represent security to the American household. The father of the frozen food industry was the naturalist Clarence Birdseye. While on a biological expedition to Labrador, Birdseye observed that fish frozen at subzero temperatures maintained its freshness and taste when thawed. For more than a decade, Birdseye worked to develop a quick freezing process that would maintain the cellular structure of food. In 1929 Birdseye's patents and equipment were purchased by what is now General Foods Corporation. A year later, a twenty-six-item line of frozen foods went to test market at ten stores in Springfield, Massachusetts. Thus a new industry was born.

Most packages on the market between 1920 and 1940 were made of paper, paperboard, and tinplate until cellophane was commercialized. The idea of taking advantage of the shrinkable properties of films originated in France in 1936 with the use of natural latex to pack perishable foods. Rubber hydrochloride (Pliofilm) was introduced as a packaging film in the same year in the United States and calendered polyvinyl chloride was introduced in Germany in 1937. In 1940, nylon as a film was introduced in Germany. Aluminum foil, used as early as 1913 for identification leg bands for racing pigeons, was beginning to show its potential. In 1921 the first aluminum-foil-laminated paperboard folding carton was produced. Successful letterpress printing on foil was developed about the same time.

Household foil was marketed in the late 1920s, and by 1929 aluminum foil accounted for 11 percent of all the metal foil produced. In 1931, aluminum foil was packaged in different sizes and thicknesses, in both rolls and sheets, as an institutional wrap primarily for use by hotel, restaurant, and hospital kitchens. Aluminum's share of foil production had rocketed to more than 50 percent by 1932. In 1937 a brewery ordered 100 million aluminum-foil bottle labels, the largest order ever received. In 1938, 56 percent of all foil produced was aluminum.

The first heat-sealing foil was developed in 1938, and many attractive new packages and labels were seen at the trade shows in 1938–39. But aluminum in all forms was soon drafted for war use, and its fullest potential in packaging was not realized until after World War II.

Another important development in packaging materials occurred with the discovery in Britain of polyethylene by Imperial Chemical Industries, Ltd. in 1933.

When America became involved in World War II, the Depression-crippled economy was still sputtering along slowly. Four million people were unemployed and 7.5 million others were earning less than the legal minimum wage of 40 cents per hour. Material shortages soon plagued the population. Inks, paperboard, tin, aluminum, and other packaging materials were either in short supply or their production was turned over to the war effort. For a time, an empty tube of toothpaste had to be returned to the storekeeper before a new one could be bought. But human ingenuity prevailed. When the Japanese swept into the East Indies, cutting off natural rubber imports from the Far East, synthetic styrene-butadiene rubber production accelerated. The development of polyvinyl chloride (PVC) was speeded up during the war when it was found that certain types of PVC could be used in place of rubber for electrical insulation. "Saran," polyvinylidene chloride copolymer film, was hustled out of the test tube into limited production by the Dow Chemical Company during the war to meet military requirements for a film that would offer superior barrier properties. It was used to package precision parts such as guns and aircraft engines.

Although DuPont's Teflon was discovered before the war, it did not make its mark until later. In the fission of Uranium 235, materials were needed with sufficient chemical resistance to handle such highly corrosive compounds as uranium hexa-flouride. Teflon was the answer, and after the war it was released for general industrial use.

The effect of World War II on aluminum foil was to establish it as a major packaging material and as a major product within the aluminum industry. During the war it was used in mostly military packaging to prevent damage to contents by moisture, vermin, and heat as electrical capacitors and insulation and as antiradar chaff, which was dropped from planes on bombing missions, as a radar shield. Following the war, aluminum for commercial use was available in large quantities. Foil packages and products soon became standard items in retail stores.

After World War II, factories that were fulfilling war contracts switched over to consumer products. The tremendous capacity to produce aluminum could not be justified by civilian demand for at least five years. "Reynolds Wrap" was test-marketed in 1947 and introduced by the Reynolds Metals Company in 1948.

Fig. A.12 *World War II grocery bag. Paper grocery bags were manufactured in long lengths in an effort to conserve paper during the war. Courtesy, the Collections of the Library of Congress.*

The post–World War II boom continued by the start of the Korean War in June 1950. Later in the 1950s fast-food franchises spread across the nation, and convenience packaging was introduced. Both product protection and consumer service was integrated in packages such as cake mixes and gravy preparations. About 1955, the TV dinner was introduced, and this was followed by other novel food and drug packages. In 1965, Elspeth Huxley wrote in *Brave New Victuals*, "You cannot sell a blemished apple in the supermarket, but you can sell a tasteless one." The era of total marketing had begun!

The pace of new packaging innovations has left industry the victim of unplanned obsolescence. These new developments occurring at a seemingly endless, non-stop pace often play havoc with the well-laid plans of management. New materials replace traditional ones; new shapes are constantly introduced. We have reached the time when the consumer perceives that the package *is* the product. After all, when cheese comes in a can and shoe polish is sold with an applicator and polishing cloth, we are in what Daniel J. Boorstin called "the packaged world."

INDEX

Page numbers in **boldface** refer to material in figures or tables.